SpringerBriefs in Statistics

For further volumes:
http://www.springer.com/series/8921

Matthias J. Fischer

Generalized Hyperbolic
Secant Distributions

With Applications to Finance

 Springer

Matthias J. Fischer
Department of Statistics and Econometrics
Friedrich-Alexander-Universität
   Erlangen-Nürnberg
Nuremberg
Germany

ISSN 2191-544X   ISSN 2191-5458   (electronic)
ISBN 978-3-642-45137-9   ISBN 978-3-642-45138-6   (eBook)
DOI 10.1007/978-3-642-45138-6
Springer Heidelberg New York Dordrecht London

Library of Congress Control Number: 2013956323

Mathematics Subject Classification (2010): 62E15, 62P20, 91G70, 91B70, 91B84

Printed on acid-free paper

Springer is part of Springer Science+Business Media (www.springer.com)

Preface

The standard normal distribution dates back to a pamphlet of de Moivre dated 12 November 1733. Further improvements were given by Laplace in 1774. The work of Gauss in 1809 and 1816 established techniques based on the normal distribution, which became standard during the nineteenth century. For both theoretical and practical reasons, the normal distribution is probably the most important distribution, not only in statistics. However, as mentioned by Chew (1968), "other probability distributions than the normal distribution may be more convenient mathematically to serve as model for the observations." Among the symmetrical distributions with an infinite domain, the most popular alternative to the normal is the logistic distribution which was already used by Verhulst (1845) in economic and demographic studies as well as the Laplace or the double exponential distribution which had its origin in 1774, where Laplace presented his first law of error. Its "two-piece nature" and its lack of differentiability at zero make the Laplace distribution sometimes unattractive and inconvenient. Occasionally, the Cauchy distribution is used, noting tails are so heavy that the mean and standard deviation as well as all higher moments are undefined. Surprisingly, one distribution avoided attracting attention, although already Manoukian and Nadeau (1988) had stated that

> ...the hyperbolic-secant distribution, which has not recieved sufficient attention in the published literature, and may be useful for students and practitioners.

During the last few years, however, several generalizations of the hyperbolic secant distribution have become popular in the context of financial return data because of its excellent fit. Nearly all of them are summarized within this Springer Brief.

Nürnberg, March 2013   Matthias J. Fischer

References

Manoukian, E.B., Nadeau, P.: A note on the hyperbolic-secant distribution. Am. Stat. **42**(1), 77–79 (1988)

Verhulst, P.F.: Recherches mathématiques sur la loi d'accroissement de la population [Mathematical Researches into the Law of Population Growth Increase]. Nouveaux Mémoires de l'Académie Royale des Sciences et Belles-Lettres de Bruxelles **18**, 1–42 (1845) http://gdz.sub.uni-goettingen.de/dms/load/img/?PPN=PPN129323640_0018&DMDID=dmdlog7. Retrieved 18 Feb 2013

Victor, C.: Some useful alternatives to the normal distribution. Am. St. **22**(3), 22–24 (1968)

Contents

Chapter 1
Hyperbolic Secant Distributions

Abstract The hyperbolic secant distribution (HSD) has its origin in Fisher [1], Dodd [2], Roa [3] and Perks [4]. Additional properties are developed by Talacko [6–8]. It is symmetric and bell-shaped like the Gaussian distribution but has slightly heavier tails. However, in contrast, both probability density function, cumulative density function and quantile function, admit simple and closed-form expressions, which makes it appealing from a practical and a theoretical point of view. In particular, HSD can be used as starting distribution to obtain generalized distribution systems which exhibit skewness and heavier (or lighter) tails.

Keywords Definition and properties · Characterizing functions · Parameter estimation

1.1 Preliminary Functions

Before the hyperbolic secant distribution (HSD) is introduced, we provide some useful results on specific functions which are intimately connected to the understanding and the derivations of the following results, and which can be skipped by the familiar reader.

Fundamental to the later developments are the so-called hyperbolic functions, in particular *hyperbolic cosine* and *hyperbolic sine* functions which are defined by

$$\cosh(x) = 0.5\left(e^x + e^{-x}\right) \quad \text{and} \quad \sinh(x) = 0.5\left(e^x - e^{-x}\right).$$

Their inverse functions are given by

$$\operatorname{acosh}(x) = \cosh^{-1}(x) = \ln(x + \sqrt{x^2 - 1}) \quad \text{and} \quad \operatorname{asinh}(x) = \ln(x + \sqrt{x^2 + 1})$$

with derivatives

M. J. Fischer, *Generalized Hyperbolic Secant Distributions*,
SpringerBriefs in Statistics, DOI: 10.1007/978-3-642-45138-6_1,
© The Author(s) 2014

$$[\mathrm{acosh}(x)]' = \frac{1}{\sqrt{x^2 - 1}} \quad \text{and} \quad [\mathrm{asinh}(x)]' = \frac{1}{\sqrt{x^2 + 1}}.$$

In addition, its reciprocal counterparts, *hyperbolic secant* and *hyperbolic cosecant* are given by

$$\mathrm{sech}(x) = \frac{1}{\cosh(x)} \quad \text{and} \quad \mathrm{csch}(x) = \frac{1}{\sinh(x)}.$$

Various series and integral representations are summarized in Gradshteyn and Ryzhik [5], e.g., the following power series representation of $\mathrm{sech}(x)$ for $x \geq 0$:

$$\mathrm{sech}(x) = \frac{2e^{-x}}{1 - (-1)e^{-2x}} = 2e^{-x} \sum_{k=0}^{\infty} \left[(-1)e^{-2x} \right]^k = 2 \sum_{k=0}^{\infty} (-1)^k e^{-(2k+1)x}. \tag{1.1}$$

According to Gradshteyn and Ryzhik [5], 1.422.1

$$\mathrm{sech}(\pi x/2) = \frac{4}{\pi} \sum_{k=1}^{\infty} (-1)^{k+1} \cdot \frac{2k - 1}{(2k - 1)^2 + x^2} \tag{1.2}$$

and, from Gradshteyn and Ryzhik [5], 1.431.4,

$$\mathrm{sech}(x) = \prod_{n=0}^{\infty} \left(1 + \frac{4x^2}{(2n + 1)^2 \pi^2} \right)^{-1}. \tag{1.3}$$

Hyperbolic functions are strongly related to trigonometric functions, e.g.,

$$\cosh(\mathbf{i}x) = \cos(x) \quad \text{and} \quad -\mathbf{i}\sinh(\mathbf{i}x) = \sin(x). \tag{1.4}$$

Occasionally, we make use of the *cosecans* function $\csc(x) = 1/\sin(x)$ and of the inverse *tangens* and *cotangens* function, i.e., $\arctan(x)$ and $\mathrm{arccot}(x)$ which are connected through the relation

$$\mathrm{arccot}(x) = \frac{\pi}{2} - \arctan(x). \quad \text{Besides,} \quad \frac{\pi}{2} - \arctan(x) = \arctan(x^{-1}) \tag{1.5}$$

for $0 < x < 1$. Recall that their derivatives are

$$\arctan'(x) = \frac{1}{1 + x^2}, \quad \mathrm{arccot}'(x) = -\frac{1}{1 + x^2}. \tag{1.6}$$

1.2 Definition

The HSD has its origin in Fisher [1], Dodd [2], Roa [3] and Perks [4]. Some of the properties are also derived by Talacko [6–8] and Lai [9]. Sections in textbooks dealing with HSDs can be found, for example, in Johnson and Kotz [10], Manoukian [11] and Manoukian and Nadeau [12]. For a brief contribution on possible generalizations we refer to Fischer [13].

Formally, a random variable $X = \ln |Y_1/Y_2|$, where Y_1, Y_2 are independent standard normal variables, is said to follow a hyperbolic secant or reciprocal of the hyperbolic cosine distribution. The corresponding density derives as follows: First, consider the distribution of $Z = Y_1/Y_2$. From Mood et al. [14], Eq. (28) in theorem 8 we conclude that

$$
\begin{aligned}
f_Z(z) &= \int_{-\infty}^{\infty} |y_1| \phi(z y_1) \phi(y_1) dy_1 \\
&= \int_{-\infty}^{\infty} |y_1| \frac{1}{2\pi} \exp\left(-0.5(y_1 \sqrt{z^2+1})^2\right) dy_1 \\
&= \frac{1}{\sqrt{2\pi(z^2+1)}} \mathbb{E}\left(|Y_1/\sqrt{z^2+1}|\right) = \frac{1}{\sqrt{2\pi(z^2+1)}} \sqrt{\frac{2}{\pi}} \\
&= \frac{1}{\pi(z^2+1)}, \quad z \in \mathbb{R}
\end{aligned}
$$

i.e., Z follows a Cauchy distribution. Second, $V = |Z|$ has a half-Cauchy distribution with density

$$
f_V(v) = \frac{2}{\pi(v^2+1)}, \quad v \geq 0.
$$

Finally, the density of $X = \ln(V)$ reads as

$$
\begin{aligned}
f(x) \equiv f_X(x) &= f_V(e^x) \cdot e^x \\
&= \frac{2e^x}{\pi(e^{2x}+1)} = \frac{2}{\pi(e^{-x}+e^x)} = \frac{1}{\pi \cosh(x)}.
\end{aligned} \tag{1.7}
$$

Obviously, the density is symmetrical around zero, i.e., $f(-x) = f(x)$ and has mode at zero with $f_X(0) = 1/\pi$. Using (1.6), (1.7) and substituting $u \equiv e^v$, the cumulative distribution function of X is

$$
F(x) = \int_{-\infty}^{x} \frac{2}{\pi} \cdot \frac{e^v}{1+(e^v)^2} dv = \frac{2}{\pi} \int_{0}^{e^x} \cdot \frac{1}{1+u^2} du = \frac{2}{\pi} \cdot \arctan(e^x). \tag{1.8}
$$

Consequently, the HS inverse function derives as

$$
F^{-1}(p) = \ln\left(\tan\left(\frac{\pi}{2} \cdot p\right)\right). \tag{1.9}
$$

and the survival or tail function by

$$S(x) = 1 - F(x) = 1 - \frac{2}{\pi} \cdot \arctan(e^x).$$

Random numbers are easily generated using either (1.9) or the representation $X = \ln|Y_1/Y_2|$ from above. Making use of these representations, Kravchuk [15] proposed a location rank test based on the HSD which is more robust to distributional misspecifications.

1.3 Properties

1. Characteristic function, moment-generating function and moments: The *characteristic function* of an HSD is

$$\mathscr{C}(t) = \mathbb{E}(e^{itX}) = \int_{-\infty}^{\infty} e^{itx} \frac{1}{\pi} \cdot \mathrm{sech}(x)dx$$

$$\stackrel{(1.1)}{=} \frac{2}{\pi} \int_{-\infty}^{\infty} e^{itx} \sum_{k=0}^{\infty} (-1)^k e^{-(2k+1)|x|}dx$$

$$= \frac{4}{\pi} \sum_{k=0}^{\infty} \frac{(-1)^k}{2k+1} \int_{-\infty}^{\infty} e^{itx} \frac{(2k+1)}{2} e^{-(2k+1)|x|}dx$$

$$= \frac{4}{\pi} \sum_{k=0}^{\infty} \frac{(-1)^k}{2k+1} \cdot \frac{1}{1 + \frac{t^2}{(2k+1)^2}} = \frac{4}{\pi} \sum_{k=1}^{\infty} (-1)^{k+1} \cdot \frac{2k-1}{(2k-1)^2 + t^2}$$

$$\stackrel{(1.2)}{=} \mathrm{sech}(\pi t/2). \tag{1.10}$$

In addition, the *moment-generating function* (mgf) of X exists and—using (1.4) and (1.10)—is given by

$$\mathscr{M}(t) = \mathbb{E}(e^{tX}) = \mathscr{C}(-it) = \sec(\pi t/2) = \frac{1}{\cos(\pi t/2)} \quad \text{for } |t| < 1.$$

Consequently, all moments exist, are finite, and given by $\mathbb{E}(X^k) = \mathscr{M}^{(k)}(0)$. In particular,

$$\mathbb{E}(X) = 0 = \mathbb{E}(X^i) \text{ for odd } i \geq 1,$$

$$Var(X) = \mathbb{E}(X^2) = \frac{\pi^2}{4}, \text{ i.e. } \sigma_X = \sqrt{Var(X)} = \frac{\pi}{2}$$

Table 1.1 Generating functions for different parameterizations of $X = \ln|Y_1/Y_2|$

No.	1	2	3
Reference	[28]	[29]	[16]
Function	$2X/\pi$	X	$2X$
Equation	(1.11)	(1.7)	(1.12), $\sigma = \pi$
$f(x)$	$(2\cosh(x\pi/2))^{-1}$	$(\pi\cosh(x))^{-1}$	$(2\pi\cosh(x/2))^{-1}$
$F(x)$	$\frac{2}{\pi}\arctan(\exp(\pi x/2))$	$\frac{2}{\pi}\arctan(\exp(x))$	$\frac{2}{\pi}\arctan(\exp(x/2))$
$F^{-1}(p)$	$\frac{2}{\pi}\ln(\tan(\pi p/2))$	$\ln(\tan(\pi p/2))$	$2\ln(\tan(\pi p/2))$
$\mathcal{M}(t)$	$(\cos(t))^{-1}$	$(\cos(t\pi/2))^{-1}$	$(\cos(t\pi))^{-1}$
$\varphi(t)$	$(\cosh(t))^{-1}$	$(\cosh(t\pi/2))^{-1}$	$(\cosh(t\pi))^{-1}$
Var	1	$\pi^2/4$	π^2

and

$$\mathbb{E}(X^4) = \frac{5\pi^4}{16}.$$

Hence, the kurtosis coefficient (i.e., the fourth standardized moment) of an HSD calculates as

$$m_4 = \frac{E(X^4)}{(Var(X))^2} = 5,$$

indicating that the HSD has heavier tails and higher peakedness than the normal distribution ($m_3 = 3$), even than the logistic distribution ($m_3 = 4.2$). Occasionally, the standardized HSD, i.e., with zero mean and unit variance is required. For this purpose, define $Z \equiv X/(\pi/2)$ with density

$$f_Z(x) = \frac{1}{e^{x\pi/2} + e^{-x\pi/2}} = \frac{1}{2} \cdot \text{sech}(x\pi/2) = \frac{1}{2\cosh(x\pi/2)}. \qquad (1.11)$$

More generally, introducing location parameter $\mu \in \mathbb{R}$ and scale parameter $\sigma > 0$, the density in (1.11) generalizes to

$$f_{HS}(x; \mu, \sigma) \equiv \frac{1}{\sigma} \cdot f_Z\left(\frac{x-\mu}{\sigma}\right) = \frac{1}{2\sigma} \cdot \frac{1}{\cosh\left(\pi \frac{x-\mu}{2\sigma}\right)}. \qquad (1.12)$$

Please note, that $X^* = \ln(G_1/G_2)$ with independent Gamma(0.5, 1)-variables G_1, G_2 is considered occasionally in the literature, which also has hyperbolic secant law with scale parameter $\sigma = \pi$ in (1.12). This family also results as a special case of the so-called exponentially generalized Beta of the second kind (briefly: EGB2), see Bondesson [16], example 7.2.4. Table 1.1 summarizes the different parameterizations.

2. Tail behavior and ψ-function: The tail behavior of a HSD is similar like that of a logistic distribution which has tail function $1 - F(x) = \exp(-x)/(1 + \exp(-x))$. The hyperbolic secant distribution exhibits (asymptotically) log-linear tails (see also

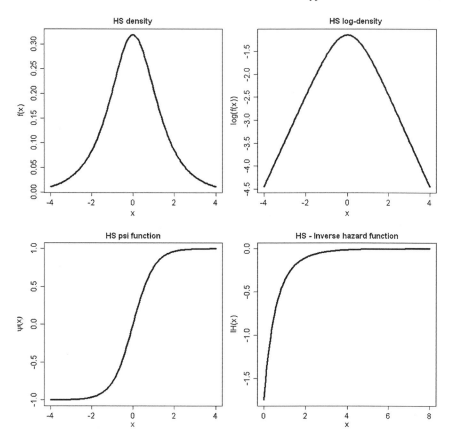

Fig. 1.1 HS distribution: Density, log-density, ψ-function, inverse hazard functio

Fig. 1.1 upper right panel) which mean that the log-density is a linear function in the left and right tails. Occasionally, such tails are termed semi-heavy (see, e.g., Sato [17]). The hyperbolic secant ψ-function[1] (see Fig. 1.1, lower left panel) is defined by

$$\psi(x) = -\frac{f'(x)}{f(x)} = \tanh(x) = \frac{e^x - e^{-x}}{e^x + e^{-x}} = \frac{e^{2x} - 1}{e^{2x} + 1}$$

approaches ± 1 for $x \to \pm\infty$.

3. Infinite divisibility: A distribution F is said to be infinitely divisible (ID) if, for each $n \geq 1$, it can be decomposed into n identical convolution factors F_n (see, e.g. Lukacs [18]). In particular, for the moment-generating function it holds that

---

[1] ψ-functions form the basic element in the context of robust statistics, in particular of robust regression, which is an alternative to least squares regression when data are contaminated with outliers or influential observations.

$\mathcal{M}(t) = (\mathcal{M}_n(t))^n$. Bondesson [16] showed that HSDs are so-called extended generalized Gamma distributions (briefly EGGC, see box below). This follows from its moment-generating function which admits the representation

$$\mathcal{M}(t) = \frac{1}{\cos(\pi t/2)} = \prod_{j=0}^{\infty} \left(1 - \frac{t}{1-2j}\right)^{-1} \left(1 - \frac{t}{1+2j}\right)^{-1}$$

as an (infinite) sum of positive and negative Gamma variables. As a consequence, HSD are self-decomposable[2] and therefore infinitely divisible (see also Pitman and Yor [19]).

Extended generalized Gamma distributions: The class of generalized Gamma convolutions (GCC) was introduced by O. Thorin [20] in 1977, as a useful tool for providing infinite divisibility of particular distributions. It is the smallest class of distributions on $\mathbb{R}_+$ that contains (positive) Gamma distributions and is closed with respect to convolution and weak limits. Noting that many distributions are limit distributions for sums of independent positive and negative Gamma variables, Thorin [21] generalized his class to so-called extended generalized Gamma distributions, briefly EGGC. For a detailed treatment on EGGCs we refer to Bondesson [16].

4. Mean-variance mixture representation: Moreover, refering to Barndorff-Nielsen et al. [22], the hyperbolic distribution admits a representation as a normal variance mixture with mixing distribution g, i.e., its density is given by

$$f(x) = \int_0^{\infty} \frac{1}{\sqrt{2\pi}} \exp(-0.5x^2/\sigma)g(\sigma)d\sigma$$

where $g(\cdot)$ denotes the density of a random variable V on $(0, \infty)$ which has moment-generating function

$$\mathcal{M}_V(t) = \prod_{i=0}^{\infty} \left(1 - \frac{2t}{(1/2+i)^2}\right)^{-1} = \frac{1}{\cos(\pi\sqrt{2t})}.$$

The variable V also appears in the first hitting time distribution for Brownian motion (see, e.g. Barndorff-Nielsen et al. [22], theorem 4.1).

5. Maximum domain of attraction: Assume now that $X_1, \ldots, X_n$ are mutually independent with common hyperbolic secant distribution function $F(x)$. As its variance exists, the normalized sample mean $\overline{X}_n = \frac{1}{n}(X_1 + \cdots + X_n)$ converges to

---

[2] A more restrictive concept than ID is self-decomposability (SD). A random variable is said to be self-decomposable if, for each c, $0 < c \leq 1$, we have $X \stackrel{d}{=} c \cdot X + \varepsilon_c$, where ε_c is a random variable independent of X.

a standardized normal distribution according to the central limit theorem. Similar results are available for the maximum of the sample: Under suitable normalization, the normalized distribution of $Z_n = \max\{X_1, \ldots, X_n\}$ converges to a Fréchet distribution $\Lambda(x)$, i.e.,

$$\lim_{n \to \infty} P(Z_n \le a_n x + b_n) = \lim_{n \to \infty} F^n(a_n x + b_n) = \Lambda(x). \qquad (1.13)$$

This can be shown on the basis of the inverse hazard function (see Fig. 1.1, lower right panel)

$$h(x) = \frac{1 - F(x)}{f(x)} = \cosh(x)(\pi - 2\text{atan}(\exp(x)))$$

and because $\lim_{x \to \infty} h'(x) = \lim_{x \to \infty} \sinh(x)(\pi - 2\text{atan}(\exp(x))) - 1 = 0$. Furthermore, $b_n \equiv F^{-1}(1 - 1/n)$ and $a_n = h(b_n)$ in (1.13). For a detailed treatment of that issue we refer to Embrechts et al. [23].

6. Self-reciprocality: The HSD is self-reciprocal as is the normal distribution. This means that its density f is proportional to its characteristic function $\mathscr{C}$ with proportional constant $\sqrt{2\pi}$:

$$\sqrt{2\pi} \cdot f(x; 0, \sqrt{\pi/2}) = \frac{\sqrt{2\pi}\sqrt{\pi}}{\sqrt{2\pi}\cosh(\sqrt{\pi}x/\sqrt{2})} = \frac{1}{\cosh(\sqrt{\pi}x/\sqrt{2})}$$
$$= \mathscr{C}(x; 0, \sqrt{\pi/2}).$$

7. Entropy: Entropy as a concept dates back to the works of Clausius in 1850 and of Boltzmann around 1870, who gave entropy a statistical meaning and related it to statistical mechanics. Next, the concept of entropy was evolved by Gibbs and Von Neumann in quantum mechanics, and was reintroduced in information theory by Shannon [24] in 1948. Information entropy is a purely probabilistic concept and is regarded as a measure of the uncertainty related to a random variable X. Given a continuous random variable X with (existing) density $f_X(x)$, the corresponding differential or Boltzmann-Gibbs-Shannon (BGS) entropy is given by

$$H(X) = H_f(X) \equiv -\int_{-\infty}^{\infty} f(x) \ln(f_X(x)) dx. \qquad (1.14)$$

If X follows a Gaussian distribution, $H(X) = 0.5 \ln(2\pi e)$. Plugging the HS density (1.7) into Eq. (1.14),

$$H(X) = \ln(\pi) + \ln(2).$$

The maximum entropy (MaxEnt) approach, established by Jaynes [25], [26], essentially relies in finding the most suitable probability distribution under the available information. According to Jaynes [25], the resulted maximum entropy distribution is the "least biased estimate possible on the given information." More formally, assume that g denotes another density and let $\mathbb{E}_g(\cdot)$ denote the expectation operator with respect to g. Further, let $\kappa = \kappa(\varepsilon)$ be a $k \times 1$ moment function for some finite number k. The MaxEnt principle recovers a least biased f by maximizing (1.14) subject to the data-consistent moment restriction

$$\mathbb{E}_f(\kappa) = \mathbb{E}_g(\kappa). \tag{1.15}$$

The solution of this problem is commonly termed as MaxEnt density $f(\cdot, \lambda_0)$ with representation given by

$$f(\varepsilon; \lambda) = \frac{1}{C(\lambda)} \exp(-\lambda' \kappa(\varepsilon)), \quad \varepsilon \in \mathbb{R}, \tag{1.16}$$

with the $k \times 1$ (parameter) vector λ and normalizing constant $C(\lambda)$, evaluated at $\lambda = \lambda_0$ which in turn corresponds to the Lagrange multiplier vector that causes (1.16) to satisfy the moment-condition (1.15). For instance, the (standard) Gaussian density solves the optimization problem

max $H(X)$ subject to restriction (1.15), where $\kappa(\varepsilon) = \varepsilon^2$, $\lambda = 1$ and $C = \sqrt{2\pi}$.

This means that the normal distribution results as MaxEnt distribution when both mean and variance are given, Similarly, the HS distribution solves the problem

max $H(X)$ subject to restriction (1.15), where $\kappa(\varepsilon) = \ln\{\cosh(\varepsilon)\}$, $\lambda = 1$ and $C = \pi$,

i.e., also results as MaxEnt distribution under a different moment condition.

8. Relations to other distributions: As seen in Sect. 1.2, the HSD basically arises from a Cauchy or the ratio of two independent Gaussian distributions. There is also an interesting relation to the logistic distribution (see, Talacko [7]). Recall that the HS density is

$$f(x) = C_1 \cdot \operatorname{sech}(x) \text{ with } C_1 = \frac{1}{\pi}.$$

Squaring this density (up to the normalizing constant), the classical logistic distribution appears:

$$g(x) = C_2 \cdot \operatorname{sech}^2(x) \text{ with } C_2 = \left[\int_{-\infty}^{\infty} \operatorname{sech}^2(x)dx\right]^{-1} = \frac{1}{2}.$$

More generally, Fisher [27] introduces the so-called z-distribution with density

$$h(x) = C_n \cdot \operatorname{sech}^n(x), \text{ where } C_n = \frac{1}{2^{n-1} B(n/2, n/2)}.$$

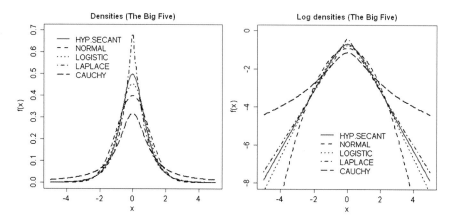

Fig. 1.2 Big 5 distributions: density and log-density

For a detailed treatment we refer to Barndorff-Nielsen et al. [22]. The z-distribution admits the stochastic representation $\frac{1}{2}\ln(F_{n:n})$, where $F_{n:n}$ denotes the classical F-distribution with (n, n) degrees of freedom.

Finally, we conclude with Fig. 1.2 which compares both the density and the log-density of the "big five" distributions on the real line, namely Normal, Logistic, hyperbolic secant, Laplace (i.e., double exponential), and Cauchy distribution. Except for the Cauchy distributions for which the variance does not exist, all variances are normalized to one.

1.4 Parameter Estimation

Assume that $X_1, ...X_n$ is an *iid* random sample from a hyperbolic secant density with unknown parameter vector $\theta = (\mu, \sigma)'$ as in (1.12). Typically, estimators $\widehat{\theta}$ of θ are obtained by the *method of moments* (MM, see Mood et al. [14], Chap. 2, for instance) or by the Maximum Likelihood method (ML, see Greene [30], Chap. 17, for instance). The main idea of the method of moments is to equate the first two moments (around zero) of the HS variable and the corresponding sample moments, i.e.,

$$\mathbb{E}(X) = \mu \stackrel{!}{=} \overline{X_n} \text{ and } \mathbb{E}(X^2) = \mu^2 + \sigma^2 \stackrel{!}{=} \overline{X_n^2}.$$

Solving for the unknown parameters μ and σ, the following MM estimators result:

$$\widehat{\mu}_{MM} = \overline{X_n}, \quad \widehat{\sigma}_{MM} = \sqrt{\overline{X_n^2} - \left(\overline{X_n}\right)^2}.$$

Alternatively, the *Maximum Likelihood* (ML) estimator of θ maximizes the log Likelihood function which in turn is defined as the logarithm of the joint density $f_{X_1,...,X_n}(x_1, ..x_n; \mu, \sigma)$ of the random sample. In case of an HS density, the log Likelihood function calculates as follows:

$$LL(\theta) \equiv \ln f_{X_1,...,X_n}(x_1, ..x_n; \theta) = \ln \prod_{i=1}^{n} f_{X_i}(x_i; \mu, \sigma)$$

$$= -n \ln(2\sigma) - \sum_{i=1}^{n} \ln \left\{ \cosh \left(\pi \frac{x_i - \mu}{2\sigma} \right) \right\}.$$

Consequently, $\widehat{\theta}_{ML}$ satisfies the so-called likelihood equation

$$\frac{\partial LL(\theta)}{\partial \theta} = \left(\frac{\partial LL(\theta)}{\partial \mu}, \frac{\partial LL(\theta)}{\partial \sigma} \right)' = 0$$

with

$$\frac{\partial LL(\theta)}{\partial \mu} = \frac{\sinh \left(\frac{\pi (x-\mu)}{2\sigma} \right) \pi}{2\sigma \cosh \left(\frac{\pi (x-\mu)}{2\sigma} \right)} \quad \text{and}$$

$$\frac{\partial LL(\theta)}{\partial \sigma} = \frac{-2\sigma \cosh \left(\frac{\pi (x-\mu)}{2\sigma} \right) + \sinh \left(\frac{\pi (x-\mu)}{2\sigma} \right) \pi x - \sinh \left(\frac{\pi (x-\mu)}{2\sigma} \right) \pi \mu}{2\sigma^2 \cosh \left(\frac{\pi (x-\mu)}{2\sigma} \right)},$$

and guarantees that $\left. \frac{\partial^2 LL(\theta)}{\partial \theta^2} \right|_{\theta=\widehat{\theta}_{ML}} < 0$. Under certain regularity conditions it is known that, if θ_0 is the true and unknown parameter vector, the ML estimator $\widehat{\theta}_{ML}$ satisfies

$$\sqrt{N}(\widehat{\theta}_{ML} - \theta_0) \xrightarrow{d} N(0, I_N^{-1}),$$

where I_N^{-1} denotes the inverse of the so-called Fisher information matrix. For practical purposes, a simple estimator of the Fisher information matrix is the so-called BHHH or outer product of gradient (OPG) estimator which is defined as

$$\widehat{I}_N^{-1} = \left[\sum_{i=1}^{n} \widehat{g}_i \widehat{g}_i' \right]^{-1} \quad \text{with } \widehat{g}_i = \frac{\partial \ln f(x_i; \widehat{\theta}_{ML})}{\partial \widehat{\theta}_{ML}}.$$

References

1. Fisher, R.A.: On the "Probable Error" of a coefficient of correlation deduced from a small sample volume I. Metron **4**, 3–32 (1921)
2. Dodd, E.L.: The frequency law of a function of variables with given frequency laws. Ann. Math. **27**(2), 13 (1925)
3. Roa, E.: A Number Of New Generating Functions with Applications to Statistics. University of Michigan, Thesis (1924)
4. Perks, W.: On some experiments in the graduation of mortality statistics. J. Inst. Actuaries **63**, 12–57 (1932)
5. Gradshteyn, I.S., Ryzhik, I.M.: Table of Integrals Series and Products. Academic Press, San Diego (2000)
6. TTalacko, J.: About some symmetrical distributions from the Perks family of functions. Ann. Math. Stat. **22**, 606–607 (1951)
7. Tacko, J.: Perks distribution and their role in the theory of Wiener's stochastic variables. Trabajos de Estatistica. **17**, 159–174 (1956)
8. Talacko, J.: A note about a family of Perks distribution, Sankhya. Indian J. Stat. **20**(3,4), 323–328 (1958)
9. Lai, C.D.: A note on a family of hyperbolic secant distribution. J. Nat. Chiao Tung Univ. **5**, 73–76 (1978)
10. Johnson, N.L., Kotz, S., Balakrishnan, N.: Continuous univariate distributions, vol. 2. John Wiley & Sons, Chicester (1995)
11. Manoukian, E.B.: Modern Concepts and Theorems of Mathematical Statistics. Springer, New York (1986)
12. Manoukian, E.B., Nadeau, P.: A note on the hyperbolic-secant distribution. Am. Stat. **42**(1), 77–79 (1988)
13. Fischer, M.: Hyperbolic secant distributions and generalizations. In: Lovric, M. (ed.) International encyclopedia of statistical science, pp. 293–294. Springer, Heidelberg (2010)
14. Mood, A.M., Graybill, F.A., Boes, D.C.: Introduction to the Theory of Statistics, 3rd edn. McGraw-Hill International Editions Statistics Series, Singapore (1974)
15. Kravchuk, O.Y.: Rank test of location optimal for hyperbolic secant distribution. Commun. Stat. Theory Methods **34**(7), 1617–1630 (2003)
16. Bondesson, L.: Generalized Gamma Convolutions and related classes of infinite divisibility. Lecture Notes in Statistics No. 76. Springer, Heidelberg (1992)
17. Sato, K.I.: Lévy Processes and Infinitely Divisible Distributions. Cambridge University Press, Cambridge (1999)
18. Lukacs, E.: Characteristic Functions, 2nd edn. Griffin, London (1970)
19. Pitman, J., Yor, M.: Infinitely divisible laws associated with hyperbolic functions. Can. J. Math. **55**(2), 292–330 (2003)
20. Thorin, O.: On the infinite divisibility of the Pareto distribution. Scand. Actuar. J. **4**, 31–40 (1977)
21. Thorin, O.: An extension of the notion of a generalized Γ-convolution. Scand. Actuar. J. **48**, 141–149 (1978)
22. Barndorff-Nielsen, O.E., Kent, J., Sœrensen, B.: Normal variance-mean mixtures and z distributions. Inte. Stat. Rev. **50**, 145–159 (1982)
23. Embrechts, P., Klüppelberg, C., Mikosch, T.: Modelling extreme events for insurance and finance. Springer, Berlin (1997)
24. Shannon, C.E.: The mathematical theory of communication. Bell Syst.Tech. J. **27**, 379–423 (1948) (Shannon, C.E and Weaver, W.: The Mathematical Theory of Communication. University of Illinois Press, Urbana, pp. 3–91 (1949) Reprinted)
25. Jaynes, E.T.: Information theory and statistical mechanics. Phys. Rev. **106**, 620–630 (1957)
26. Jaynes, E.T.: Information theory and statistical mechanics II, Phys. Rev. **108**, 171–190 (1957)
27. Fisher, R.A.: On a distribution yielding the error functions of several well-known statistics. Proc. Int. Math. Congr. **2**, 805–813 (1924)

28. Baten, W.D.: The probability law for the sum of n independent variables, each subject to the law $(1/(2h)) \operatorname{sech} (\pi x/(2h))$. Bull. Am. Math. Soc. **40**, 284–290 (1934)
29. Vaughan, D.C.: The generalized secant hyperbolic family and its properties. Communi. Stat. Theory Methods **31**(2), 219–238 (2002)
30. Greene, W.H.: Econometric analysis. Prentice Hall, NeW Jersey (2003)

Chapter 2
The GSH Distribution Family and Skew Versions

Abstract The generalized secant hyperbolic (GSH) distribution denotes a popular symmetric subclass of Perk's family which was already introduced in 1932. It allows for any kurtosis higher than 1.8 and, hence, admits both thin and fat tail behavior. Under a slightly different parameterization, the GSH family was re-examined by [1] who also derived additional properties. Based on the GSH family, there are three different proposals in the literature—related to Fischer and Vaughan [2], Fischer [3], and Vaughan [4]—how to additionally introduce skewness which are discussed within this chapter.

Keywords Definition and properties · Perk's distribution · Scale parameter split · Esscher transformation · Vaughan's skew version

2.1 Perk's Distribution Family

Already in 1932, the British actuary Wilfred Perks [5]—being interested in general functions for graduating life-table data—introduced a large class of probability densities of the form

$$f(x) = \frac{a_0 + a_1 e^{-x} + a_2 e^{-2x} + \cdots + a_m e^{-mx}}{b_0 + b_1 e^{-x} + b_2 e^{-2x} + \cdots + b_n e^{-nx}} \tag{2.1}$$

with parameters $a_0, a_1, \ldots, a_m, b_0, b_1, \ldots, b_n$ such that f is actually a probability density. Setting $m = 1, a_0 = 0, a_1 = 1$ and $n = 2, b_0 = 1, b_1 = 0, b_2 = 1$, Eq. (2.1) reduces to hyperbolic secant distribution:

$$f(x) = \frac{2}{\pi} \cdot \frac{e^{-x}}{1 + e^{-2x}} = \frac{1}{\pi} \cdot \frac{1}{\cosh(x)}, \quad x \in \mathbb{R}.$$

M. J. Fischer, *Generalized Hyperbolic Secant Distributions*,
SpringerBriefs in Statistics, DOI: 10.1007/978-3-642-45138-6_2,
© The Author(s) 2014

For $b_0 = b_2 = 2$, the logistic distribution is recovered. Slightly more generally, Talacko [6] discussed specific distribution families with $m = 1$, $a_0 = 0$ and $n = 2$, $b_0 = b_2$, i.e., densities of the form

$$f(x) = \frac{a_1 e^{-x}}{b_0 + b_1 e^{-x} + b_0 e^{-2x}} = \frac{c}{e^x + k + e^{-x}} = \frac{c e^x}{e^{2x} + k e^x + 1}, \qquad x \in \mathbb{R}$$

(2.2)

where $c \equiv a_1 / b_0$ is a normalizing constant and $k \equiv b_1 / b_0 > -2$ makes sure that (2.2) is actually a density. For $-2 < k \le 2$ but $k \ne 0$ replace k in (2.2) by $2 \cos(\lambda)$ with $0 \le \lambda < \pi$. Talacko [6] calculated the corresponding characteristic functions as follows:

$$\mathscr{C}(t) = \mathbb{E}(e^{itX}) = c \int_{-\infty}^{\infty} \frac{e^{itx} dx}{e^x + 2\cos(\lambda) + e^{-x}} = c \int_{-\infty}^{\infty} \frac{e^{(it+1)x} dx}{e^{2x} + 2\cos(\lambda)e^x + 1}$$

$$= c \int_{-\infty}^{\infty} \frac{e^{(it+1)x} dx}{(e^x + e^{i\lambda})(e^x + e^{-i\lambda})} = c \int_C \frac{e^{(it+1)z} dz}{(e^z + e^{i\lambda})(e^z + e^{-i\lambda})}$$

$$= c \cdot \frac{\pi}{\sin(\lambda)} \cdot \frac{\sinh(\lambda t)}{\sinh(\pi t)} = \frac{\pi}{\lambda} \cdot \frac{\sinh(\lambda t)}{\sinh(\pi t)}.$$

Note that from $\lim_{t \to 0} \mathscr{C}(t) = 1$ we concluded that $c = \frac{\sin(\lambda)}{\lambda}$. For $k > 2$, replace λ by $i\theta$, i.e. k by $\cos(i\theta) = \cosh(\theta)$ in order to obtain with a similar calculation

$$\mathscr{C}(t) = \frac{\pi}{\theta} \cdot \frac{\sin(\theta t)}{\sinh(\pi t)} \quad \text{and} \quad c = \frac{\sinh(\theta)}{\theta}.$$

It took about 50 years until Talacko's generalized secant hyperbolic (GSH) distribution was re-examined by Vaughan [1] under the slightly different parameterization

$$k = k(\eta) = \begin{cases} \cos(\eta), & -\pi < \eta \le 0, \\ \cosh(\eta), & \eta \ge 0 \end{cases}$$

and with scaling constant $c_2 = c_2(\eta)$ such that zero mean and unit variance is achieved:

$$f(x; \eta) = c_1(\eta) \cdot \frac{\exp(c_2(\eta)x)}{\exp(2c_2(\eta)x) + 2a(\eta)\exp(c_2(\eta)x) + 1}$$

(2.3)

$$= \frac{c_1(\eta)}{2(\cosh(c_2(\eta)x) + a(\eta))}$$

(2.4)

with

$$a(\eta) = \cos(\eta), \quad c_2(\eta) = \sqrt{\frac{\pi^2 - \eta^2}{3}}, \quad c_1(\eta) = \frac{\sin(\eta)}{\eta} \cdot c_2(\eta) \quad \text{for } \eta \in (-\pi, 0],$$

$$a(\eta) = \cosh(\eta), \quad c_2(\eta) = \sqrt{\frac{\pi^2 + \eta^2}{3}}, \quad c_1(\eta) = \frac{\sinh(\eta)}{\eta} \cdot c_2(\eta) \quad \text{for } \eta > 0.$$

Vaughan [1] also derived the cumulative distribution function, given by

$$F(x; \eta) = \begin{cases} 1 + \frac{1}{\eta} \operatorname{arccot}\left(-\frac{\exp(c_2(\eta)x) + \cos(\eta)}{\sin(\eta)}\right) & \text{for } \eta \in (-\pi, 0), \\ \frac{\exp(\pi x/\sqrt{3})}{1 + \exp(\pi x/\sqrt{3})} & \text{for } \eta = 0, \\ 1 - \frac{1}{\eta} \operatorname{arccoth}\left(\frac{\exp(c_2(\eta)x) + \cosh(\eta)}{\sinh(\eta)}\right) & \text{for } \eta > 0 \end{cases}$$

and the inverse distribution function, given by

$$F^{-1}(u; \eta) = \begin{cases} \frac{1}{c_2(\eta)} \ln\left(\frac{\sin(\eta u)}{\sin(\eta(1-u))}\right) & \text{for } \eta \in (-\pi, 0), \\ \frac{\sqrt{3}}{\pi} \ln\left(\frac{u}{1-u}\right) & \text{for } \eta = 0, \\ \frac{1}{c_2(\eta)} \ln\left(\frac{\sinh(\eta u)}{\sinh(\eta(1-u))}\right) & \text{for } \eta > 0. \end{cases}$$

2.2 Properties of the GSH Family

The density from (2.3) is chosen so that the GSH variable has zero mean and unit variance, the range of the "kurtosis parameter" η is $\in (-\pi, \infty)$. Actually, Fischer and Klein [16] proved that the η is a kurtosis parameter in the sense of van Zwet [7]. The GSH distribution includes the logistic distribution ($\eta = 0$) and the hyperbolic secant distribution ($\eta = -\pi/2$) as special cases and the uniform distribution on $(-\sqrt{3}, \sqrt{3})$ as limiting case for $\eta \to \infty$. Figure 2.1 displays different densities and log-density. All densities are unimodal.

The moment-generating function also depends on η and is given by

$$\mathcal{M}(u; \eta) = \begin{cases} \frac{\pi}{\eta} \sin(u\eta/c_2(\eta)) \csc(u\pi/c_2(\eta)) & \text{for } \eta \in (-\pi, 0), \\ \sqrt{3}u \csc(\sqrt{3}u) & \text{for } \eta = 0, \\ \frac{\pi}{\eta} \sinh(u\eta/c_2(\eta)) \csc(u\pi/c_2(\eta)) & \text{for } \eta > 0. \end{cases}$$

It also satisfies (see Vaughan [1], p. 222)

$$\mathcal{M}(u; \eta) = \begin{cases} 1 + \frac{1}{2}u^2 + \frac{1}{4!} \frac{21\pi^2 - 9\eta^2}{5\pi^2 - 5\eta^2} u^4 + \mathcal{O}(u^5) & \text{for } \eta \in (-\pi, 0], \\ 1 + \frac{1}{2}u^2 + \frac{1}{4!} \frac{21\pi^2 + 9\eta^2}{5\pi^2 + 5\eta^2} u^4 + \mathcal{O}(u^5) & \text{for } \eta > 0 \end{cases}$$

establishing that $Var(X) = 1$, so that the kurtosis coefficient $m_4 = \mathbb{E}(X^4)$ is

$$m_4 = \begin{cases} \frac{21\pi^2 - 9\eta^2}{5\pi^2 - 5\eta^2} & \text{for } \eta \in (-\pi, 0], \\ \frac{21\pi^2 + 9\eta^2}{5\pi^2 + 5\eta^2} & \text{for } \eta > 0. \end{cases}$$

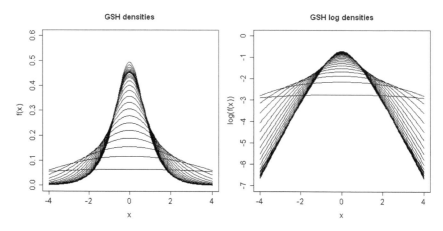

Fig. 2.1 GSH distribution: Log-density, density for different $\eta \in [-\pi, 5]$

It is readily apparent that m_4 decreases as η tends to ∞ and that $m_4 \in (1.8, \infty)$. Vaughan [1] also states that there is a unique member of the GSH family that corresponds to any given kurtosis for regular unimodal distributions:

$$\eta = -\pi \sqrt{\frac{5m_4 - 21}{5m_4 - 9}} \text{ for } m_4 \geq 4.2 \text{ and } \eta = \pi \sqrt{\frac{21 - 5m_4}{5m_4 - 9}} \text{ for } m_4 \leq 4.2.$$

In particular, when $\eta = \pi$ then $m_4 = 3$, the kurtosis of a normal distribution. Note also that if ν denotes the degrees of freedom for a Student-t distribution with a given (finite) kurtosis, then the parameter η in the GSH family with the same first four moments is $-\pi \sqrt{(9 - \nu)/(\nu + 1)}$ for $4 < \nu < 9$, 0 for $\nu = 9$ and $\pi \sqrt{(\nu - 9)/(\nu + 1)}$ for $\nu > 9$.

2.3 Introducing Skewness by Splitting the Scale Parameter

The first skew version of Vaughan's GSH distribution was proposed by Fischer and Vaughan [2]. Application to unconditional and conditional financial return models followed up with Fischer [8] and Palmitesta and Provasi [9]. The main idea of this approach is to split the scale parameter of the GSH distribution into two parameters representing the left and the right part across the expectation value. Note that this idea was already used by Fernández et al. [10] and Fernández and Steel [11] in order to design a skew Student-t distribution. Let $\mathbf{I}^+(x)$ denote the indicator function for x on $\mathbb{R}_+$ and $\mathbf{I}^-(x)$ the indicator function for x on $\mathbb{R}_-$. It follows that

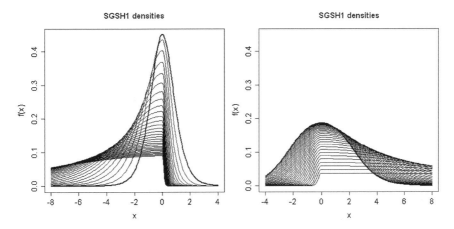

Fig. 2.2 SGSH1 distribution: Effect of γ for fixed $\eta = 0.2$ and $\gamma \in [1, 10]$ (*left panel*) and $\eta = 2.5$ and $\gamma \in [0.1, 1]$ (*right panel*)

$$f(x; \eta, \gamma) = \frac{2\gamma}{\gamma^2 + 1} \left\{ f_{GSH}(x/\gamma; \eta) \cdot \mathbf{I}^-(x) + f_{GSH}(\gamma x; \eta) \cdot \mathbf{I}^+(x) \right\}$$

$$= \frac{2c_1}{\gamma + \frac{1}{\gamma}} \cdot \left(\frac{\exp(c_2 x/\gamma) \cdot \mathbf{I}^-(x)}{\exp(2c_2 x/\gamma) + 2a \exp(c_2 x/\gamma) + 1} + \frac{\exp(c_2 \gamma x) \cdot \mathbf{I}^+(x)}{\exp(2c_2 \gamma x) + 2a \exp(c_2 \gamma x) + 1} \right)$$

is a density function which is symmetric for $\gamma = 1$, skewed to the right for $\gamma > 1$ and skewed to the left for $0 < \gamma < 1$. The corresponding distribution will be termed the *skewed GSH distribution of type I* (SGSH1) in the sequel. The effect of γ on the GSH density is illustrated in Fig. 2.2.

Following Fischer [3], both cumulative distribution function and quantile function admit closed forms, namely

$$F(x; \eta, \gamma) = \frac{2\gamma^2}{\gamma^2 + 1} \cdot \left(F_{GSH}(x/\gamma) \cdot \mathbf{I}^-(x) + \left(\frac{\gamma^2 - 1 + 2F_{GSH}(\gamma x)}{2\gamma^2} \right) \cdot \mathbf{I}^+(x) \right),$$

$$F^{-1}(x; \eta, \gamma) = \gamma F_{GSH}^{-1}\left(x \cdot \frac{\gamma^2 + 1}{2\gamma^2}; \eta\right) \mathbf{I}_A^-(x) + \frac{1}{\gamma} F_{GSH}^{-1}\left(x \cdot \frac{\gamma + 1}{2} - \frac{\gamma - 1}{2}; \eta\right) \mathbf{I}_A^+(x).$$

with

$$\mathbf{I}_A^-(x) = \begin{cases} 1, & \text{if } x < \frac{\gamma^2}{1+\gamma}, \\ 0, & \text{if } x \geq \frac{\gamma^2}{1+\gamma}. \end{cases} \quad \text{and} \quad \mathbf{I}_A^+(x) = 1 - \mathbf{I}_A^-(x).$$

Referring to Fernández and Steel [11], the power moments of an SGSH1-variable Z can be derived using the following calculation scheme:

$$\mathbb{E}(Z^r) = \mathbb{E}^+(X^r) \cdot \frac{2\gamma}{\gamma^2 + 1} \cdot \left(\gamma^{-r-1} + (-1)^r \gamma^{r+1}\right) \text{ with } \mathbb{E}^+(X^r) \equiv \int_0^\infty x^r f_{GSH}(x) dx.$$

Evidently, $\mathbb{E}^+(X^r)$ equals the r-th power moment of the GSH distribution (which can be obtained from the corresponding moment-generating function) divided by two when r is even. For odd r and $t \neq 0$, Palmitesta and Provasi [9] derive the following expression[1]:

$$\mathbb{E}^+(X^r) = \frac{c_1 \Gamma(r+1)}{2 c_2^{r+1} \sqrt{a^2 - 1}} \cdot \mathscr{L}_{r,a}$$

defining

$$\mathscr{L}_{r,a} \equiv \left[\mathscr{L}_{r+1}\left(-\frac{1}{\sqrt{a^2 - 1} + a}\right) - \mathscr{L}_{r+1}\left(\frac{1}{\sqrt{a^2 - 1} - a}\right)\right], \text{ where } \mathscr{L}_r(x) \equiv \sum_{k=1}^\infty \frac{x^k}{k^r}$$

denotes the *polylogarithmic function* (see Lewin [12]) which is defined for $x \in \mathbb{C}$ and $|x| < 1$. Consequently, the first four power moments derive as

$$\mathbb{E}(Z) = \frac{c_1 \mathscr{L}_{1,a}(1-\gamma^2)}{\gamma c_2^2 \sqrt{a^2 - 1}}, \quad \mathbb{E}(Z^2) = \frac{\gamma^4 - \gamma^2 + 1}{\gamma^2},$$

$$\mathbb{E}(Z^3) = \frac{6 c_1 \mathscr{L}_{2,a}(1 - \gamma^6 + \gamma^4 - \gamma^2)}{\gamma^3 c_2^4 \sqrt{a^2 - 1}}$$

and

$$\mathbb{E}(Z^4) = \frac{(21\pi^2 + \text{sgn}(t)9t^2)(1 + \gamma^8 - \gamma^2 - \gamma^6 + \gamma^4)}{\gamma^4(5\pi^2 + \text{sgn}(t)5\eta^2)}.$$

Using the relationship between the centered and uncentered moments (e.g., Stuart and Ord [13]), the derivation of the third and fourth standardized moments is tedious but straightforward. As expected, both η and γ determine the skewness and the kurtosis. It should also be noted that all other higher moments exist.

2.4 Introducing Skewness by Means of the Esscher Transformation

A second way to introduce skewness into Vaughan's GSH distribution was recently discussed by Fischer [3], who used the existence of the moment-generating function to construct asymmetric densities by means of the so-called *Esscher transformation*.

---

[1] We exclude the case $\eta = 0$, which corresponds to the logistic distribution and refers to Palmitesta and Provasi [9], instead.

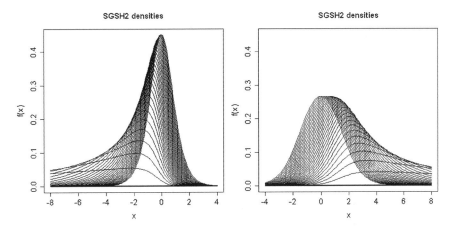

Fig. 2.3 SGSH2 distribution: h for fixed $\eta = 0.2$ and $h \in [-1, 0]$ *(left panel)*, and $\eta = 2$ and $h \in [0, 1]$ *(right panel)*

Esscher transformation: Originally, this concept was a tool in actuarial science suggested by Esscher [14], which was popularized by Gerber and Shiu [15] who applied this concept to value derivative securities. Given a random variable X with moment-generating function $\mathscr{M}_X(t)$ and density $f_X(x)$, the Esscher-transformed density with parameter h is defined by

$$f(x; h) \equiv \exp(hx) f(x) / \mathscr{M}(h). \qquad (2.5)$$

Note that if X is Gaussian, the resulting Esscher-transformed variable is again Gaussian (and thus symmetric) but with different scale and location. In contrast, Esscher-transformations of symmetric non-Gaussian densities frequently produce asymmetric distributions, where $h \neq 0$ governs the amount of skewness, and symmetry is obtained for $h = 0$.

Plugging (2.3) into (2.5), the Esscher-transformed GSH density for $h \neq 0$ derives as

$$f(x; \eta, h) \equiv \begin{cases} \dfrac{\sin(h\pi)\sin(\eta)}{\pi \sin(ht)} \cdot \dfrac{\exp((h+1)x)}{\exp(2x)+2\cos(\eta)\exp(x)+1} & \text{for } -\pi < \eta < 0, \\[2ex] \dfrac{\sin(h\pi)}{h\pi} \cdot \dfrac{\exp((h+1)x)}{\exp(2x)+2\exp(x)+1} & \text{for } \eta = 0, \\[2ex] \dfrac{\sin(h\pi)\sinh(\eta)}{\pi \sinh(h\eta)} \cdot \dfrac{\exp((h+1)x)}{\exp(2x)+2\cosh(\eta)\exp(x)+1} & \text{for } \eta > 0 \end{cases} \qquad (2.6)$$

and will be termed as *skew GSH densities* of type II, or briefly SGSH2 densities in the sequel. Examples of SGSH2 densities are plotted in Fig. 2.3.

Note that for $-\pi < \eta < 0$ and $h \neq 0$ we can derive the corresponding *moment-generating function* of a SGSH2 variable X,

$$\mathbb{E}(e^{uX}) = \int_{-\infty}^{\infty} \frac{\sin(h\pi)\sin(\eta)}{\pi \sin(ht)} \frac{\exp((h+u+1)x)}{\exp(2x) + 2\cos(\eta)\exp(x) + 1} dx$$

$$= \frac{\sin((h+u)\eta)}{\sin((h+u)\pi)} \frac{\sin(h\pi)}{\sin(h\eta)} \int_{-\infty}^{\infty} \frac{\sin((h+u)\pi)\sin(\eta)\exp(((h+u)+1)x)dx}{\pi \sin((h+u)\eta)(\exp(2x) + 2\cos(\eta)\exp(x) + 1)}.$$

$$= \frac{\sin((h+u)\eta)}{\sin((h+u)\pi)} \frac{\sin(h\pi)}{\sin(h\eta)}.$$

Similar reformulations hold for $\eta \geq 0$ and we finally arrive at

$$\mathcal{M}(u) = \begin{cases} (\sin((h+u)\eta)\sin(h\pi)) \,/\, (\sin((h+u)\pi)\sin(h\eta)) & \text{for } -\pi < \eta < 0, \\ ((h+u)\cdot\sin(h\pi)) \,/\, (h\sin((h+u)\pi)) & \text{for } \eta = 0, \\ (\sinh((h+u)\eta)\sin(h\pi)) \,/\, (\sin((h+u)\pi)\sinh(h\eta)) & \text{for } \eta > 0. \end{cases}$$

All moments of the SGSH2 distribution exist. In particular, the first four power moments are given by

$$\mathbb{E}(X) = \begin{cases} \eta\cot(h\eta) - \pi\cot(h\pi) & \text{for } -\pi < \eta < 0, \\ (1 - h\pi\cot(h\pi))/h & \text{for } \eta = 0, \\ \eta\coth(h\eta) - \pi\cot(h\pi) & \text{for } \eta > 0, \end{cases}$$

$$\mathbb{E}(X^2) = \begin{cases} \pi^2 - \eta^2 - 2\eta\pi\cot(h\eta)\cot(h\pi) + 2\pi^2\cot^2(h\pi) & \text{for } -\pi < \eta < 0, \\ \pi^2 - 2\pi/h\cdot\cot(h\pi) + 2\pi^2\cot^2(h\pi) & \text{for } \eta = 0, \\ \eta^2 + \pi^2 - 2\eta\pi\coth(h\eta)\cot(h\pi) + 2\pi^2\cot^2(h\pi) & \text{for } \eta > 0, \end{cases}$$

$$\mathbb{E}(X^3) = \begin{cases} -\eta^3\cot(h\eta) + 3\eta^2\pi\cot(h\pi) + 6\eta\pi^2\cot(h\eta)\cot^2(h\pi) + 3\eta\pi^2\cot(h\eta) \\ \quad -6\pi^3\cot^3(h\pi) - 5\pi^3\cot(h\pi) \qquad \text{for } -\pi < \eta < 0, \\ 6\pi^2/h\cdot\cot^2(h\pi) + 3\pi^2/h\cdot\cot(h\pi) - 6\pi^3\cot^3(h\pi) - 5\pi^3\cot(h\pi) \qquad \text{for } \eta = 0, \\ \eta^3\coth(h\eta) - 3\eta^2\pi\cot(h\pi) + 6\eta\pi^2\coth(h\eta)\cot^2(h\pi) + 3\eta\pi^2\coth(h\eta) \\ \quad -6\pi^3\cot^3(h\pi) - 5\pi^3\cot(h\pi) \qquad \text{for } \eta > 0, \end{cases}$$

$$\mathbb{E}(X^4) = \begin{cases} \eta^4 + 5\pi^4 - 4\eta^3\pi\coth(h\eta)\cot(h\pi) + 12\eta^2\pi^2\cot^2(h\pi) + 6\eta^2\pi^2 \\ \quad -24\eta\pi^3\coth(h\eta)\cot^3(h\pi) - 20\eta\pi^3\coth(h\eta)\cot(h\pi) \\ \quad +24\pi^4\cot^4(h\pi) + 28\pi^4\cot^2(h\pi) \qquad \text{for } -\pi < \eta < 0. \\ 5\pi^4 - 24\pi^3/h\cdot\cot^3(h\pi) - 20\pi^3/h\cot(h\pi) + 24\pi^4\cot^4(h\pi) \\ \quad +28\pi^4\cot^2(h\pi) \qquad \text{for } \eta = 0, \\ \eta^4 + 5\pi^4 + 4\eta^3\pi\cot(h\eta)\cot(h\pi) - 12\eta^2\pi^2\cot^2(h\pi) - 6\eta^2\pi^2 \\ \quad -24\eta\pi^3\cot(h\eta)\cot^3(h\pi) - 20\eta\pi^3\cot(h\eta)\cot(h\pi) + 24\pi^4\cot^4(h\pi) \\ \quad +28\pi^4\cot^2(h\pi) \qquad \text{for } \eta > 0, \end{cases}$$

Consequently, the variance of an SGSH2 variable is given by

$$Var(X) = \begin{cases} \pi^2(1 + \cot^2(h\pi)) - \eta^2(1 + \cot^2(h\eta)) & \text{for} \quad -\pi < \eta < 0, \\ \pi^2(1 + \cot^2(h\pi)) - 1/h^2, & \text{for} \quad \eta = 0, \\ \pi^2(1 + \cot^2(h\pi)) + \eta^2(1 - \coth^2(h\eta)) & \text{for} \quad \eta > 0. \end{cases}$$

2.5 Vaughan's Skew Extension

Recently, Vaughan [4] advocated skew-extended GSH (S-EGSH) distribution families as a natural generalization of the GSH representative. For constants $c > h \geq 0$, $k, \omega q > 0$ and parameters a and b satisfying either $\omega k b > a > 0$ or $0 > a > \omega k b$ Vaughan [4] discusses, e.g.,

$$f(x) = C_1 \frac{\exp(ax)}{\left[(\exp(bx) + c)^k - h^k\right]^\omega} \tag{2.7}$$

with normalizing constant

$$C_1 = bc^{-\lambda + k\omega} \left[\sum_{j=0}^{\infty} \frac{\Gamma(\omega + j)}{\Gamma(\omega)j!} \kappa^{kj} B(\lambda, k(j + \omega) - \lambda) \right]^{-1},$$

where $B(a, b)$ denotes the Beta function, $\lambda = a/b$ and $\kappa = h/c$. The original GSH family in (2.4) is recovered by setting

$$a = b = c_2(\eta), \quad c = a(\eta), \quad h = \sqrt{a(\eta)^2 - 1}, \quad k = 2, \quad w = 1$$

in (2.7). The corresponding cumulative function is given by

$$F(x) = C_1 \frac{\sum_{j=0}^{\infty} \frac{\Gamma(\omega+j)}{j!} \kappa^{kj} B_{\tau(u)}(\lambda, k(j + \omega) - \lambda)}{\sum_{j=0}^{\infty} \frac{\Gamma(\omega+j)}{j!} \kappa^{kj} B(\lambda, k(j + \omega) - \lambda)} \quad \text{for} \quad \tau(u) \equiv \frac{c}{\exp(bu) + c}$$

and where $B_u(a, b)$ denotes the incomplete Beta function. The above conditions on the parameters ensure the densities are positive, and further that the distribution has well-defined moment-generating functions, and hence all moments finite. They can be expressed in terms of the Gamma function and its derivatives. For all S-EGSH members there is a unique mode (Fig. 2.4).

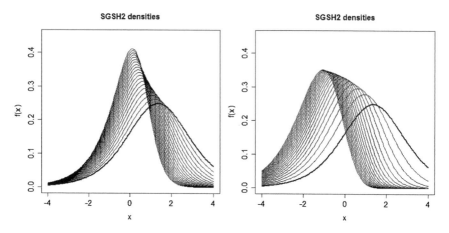

Fig. 2.4 S-EGSH distribution: Densities with $a = 1$, $C = 4$, $h = 1$, $k = 2$, $w = 1$ and $b \in [1, 2]$ (*left panel*), and $a = 1$, $b = 1$, $C = 4$, $h = 1$, $k = 2$ and $w \in [1, 6]$ (*right panel*)

References

1. Vaughan, D.C.: The generalized secant hyperbolic family and its properties. Commun. Stat. Theory Methods **31**(2), 219–238 (2002)
2. Fischer, M., Vaughan, D.C.: Classes of skewed generalized hyperbolic secant distributions. Working Paper No. 45 (unpublished). Department of Statistics and Econometrics, FAU Erlangen-Nuremberg, Nuremberg (2002)
3. Fischer, M.: A skew generalized secant hyperbolic family. Austrian J. Stat. **35**(4), 437–444 (2006)
4. Vaughan, D.C.: Modeling skewness and kurtosis with the S-EGSH. Working Paper (unpublished). Department of Mathematics, Wilfrid Laurier University, Waterloo (2011)
5. Perks, W.: On some experiments in the graduation of mortality statistics. J. Inst. Actuar. **63**, 12–57 (1932)
6. Talacko, J.: Perks' distribution and their role in the theory of Wiener's stochastic variables. Trabajos de Estatistica **17**, 159–174 (1956)
7. van Zwet, W.R.: Convex Transformations of Random Variables. Mathematical Centre Tracts No. 7, Mathematical Centre, Amsterdam (1964).
8. Fischer, M.: Skew generalized secant hyperbolic distributions: unconditional and conditional fit to asset returns. Austrian J. Stat. **33**(3), 293–304 (2004)
9. Palmitesta, P., Provasi, C.: GARCH-type models with generalized secant hyperbolic innovations. Stud. Nonlinear Dyn Econom. **8**(2), 1–17 (2004)
10. Fernández, C., Osiewalski, J., Steel, M.F.J.: Modelling and inference with v-spherical distributions. J. Am. Stat. Assoc. **90**, 1331–1340 (1995)
11. Fernández, C., Steel, M.F.J.: On Bayesian modelling of fat tails and skewness. J. Am. Stat. Assoc. **93**, 359–371 (1998)
12. Lewin, L.: Polylogarithms and Associated Functions. North-Holland, New York (1981)
13. Stuart, A., Ord, K.: Kendall's advanced theory of statistics. Stuart, A., Ord, K. (eds.) Distribution theory, vol. 1, Wiley, New York (1994)
14. Esscher, E.: On the probability function in the collective theory of risk. Skandinavisk Aktuarietidskrifi **15**, 175–195 (1932)
15. Gerber, H.U., Shiu, E.S.W.: Option pricing by Esscher transforms. Trans. Soc Actuar. **46**, 99–191 (1994)

16. Klein, I., Fischer, M.: A note on the kurtosis ordering of the generalized secant hyperbolic family. Commun. Stat.Theory Methods **37**(1), 1–7 (2008)
17. Kravchuk, O.Y.: R-estimator of location of the GSHD. Commun. Stat Simul. Comput. **35**(1), 1–18 (2006)
18. Talacko, J.: About some symmetrical distributions from the Perks' family of functions. Ann. Math. Stat. **22**, 606–607 (1951)
19. Talacko, J.: A note about a family of Perks' distribution. Sankhya: Indian J. Stat. **20**(3,4), 323–328 (1958)

Chapter 3
The NEF-GHS or Meixner Distribution Family

Abstract Another way to generalize the hyperbolic secant distribution with respect to its tail behavior is to consider the ρ-th convolution. This was originally investigated by Baten (1934) for $\rho \in \mathbb{N}$ followed by Baten's [11] generalization for arbitrary $\rho > 0$. In order to introduce skewness, it was suggested to apply the Esscher transformation afterwards. The resulting distribution is frequently termed as NEF-GHS or Meixner distribution and is the focus of this chapter.

Keywords Definition and properties · Convolution of HS distribution · Natural exponential family · Random number generation

3.1 GHS Distribution: Definition and History

Another line of research which became popular in finance has its roots in the work of Baten [3] who derived the probability density function of

$$X \equiv X_1 + \cdots + X_n,$$

where $X_1, \ldots, X_n$ are independent hyperbolic secant copies with scale parameter $\alpha > 0$ for finite $n \in \mathbb{N}$. In the limit $n = \infty$, $\overline{X}_n$ converges to a normal variable with variance 2. More generally, Harkness and Harkness [11] discuss distribution families with characteristic function

$$\mathscr{C}(t) = \operatorname{sech}(\alpha t)^\rho, \qquad \alpha > 0, \ \rho > 0 \tag{3.1}$$

which can be identified as the ρ-th convolution of a hyperbolic secant variable. This family is commonly known as *generalized hyperbolic secant* (GHS) distribution with shape parameter ρ and scale parameter α (which we set to one henceforth). The underlying GHS density follows from the inversion formula and Gradshteyn and Ryzhik ([7], 3.985.1):

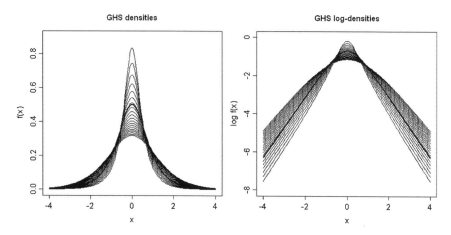

Fig. 3.1 GHS distribution: Density and log-density for $\rho \in [0.5, 3]$

$$
\begin{aligned}
f(x; \rho) &= \frac{1}{2\pi} \int_{-\infty}^{\infty} e^{-itx} \mathscr{C}(t) dt = \frac{1}{\pi} \int_0^{\infty} \cos(tx) \cdot \mathrm{sech}(t)^\rho dt \\
&= \frac{2^{\rho-2}}{\pi \Gamma(\rho)} \cdot \Gamma\left(\frac{\rho}{2} + i\frac{x}{2}\right) \Gamma\left(\frac{\rho}{2} - i\frac{x}{2}\right) \\
&= \frac{2^{\rho-2}}{\pi \Gamma(\rho)} \cdot \left|\Gamma\left(\frac{\rho}{2} + i\frac{x}{2}\right)\right|^2 \\
&= \frac{2^{\rho-2}\Gamma^2(\rho/2)}{\pi \Gamma(\rho)} \cdot \prod_{n=0}^{\infty} \left[1 + \left(\frac{x}{\rho + 2n}\right)^2\right]^{-1},
\end{aligned}
\tag{3.2}
$$

where the last equality follows from Abramowitz and Stegun ([1], p. 256). The standardized counterpart (see Fig. 3.1) with variance one is

$$
f^*(x; \rho) \equiv \frac{2^{\rho-2}\sqrt{\rho}}{\pi \Gamma(\rho)} \cdot \left|\Gamma\left(\frac{\rho}{2} + i\frac{\sqrt{\rho}\, x}{2}\right)\right|^2.
\tag{3.3}
$$

Simpler forms of the density are provided by Baten [3] and Harkness and Harkness [11] for $\rho \in \mathbb{N}$. For $\rho = 1$, the hyperbolic secant density from Chap. 1 is recovered using Gradshteyn and Ryzhik ([7], 8.332.2),

$$
|\Gamma(0.5 + ix)|^2 = \frac{\pi}{\cosh(\pi x)}.
$$

For even $\rho = 2n$ ($n \in \mathbb{N}$) in (3.2), the corresponding counterpart is

$$
|\Gamma(n + ix)|^2 = \frac{\pi P_n}{x \sinh(\pi x)} \quad \text{for} \quad n = 1, 2, \dots \quad \text{and} \quad P_n \equiv \prod_{i=1}^{n}\left[(i-1)^2 + x^2\right].
$$

3.2 GHS Distribution: Properties

1. Distribution function, moment-generating function and moments: Unfortunately, the cumulative distribution admits no closed form in general and has to be calculated numerically from the characteristic function (e.g., using Fast Fourier Transformation) or from the density (e.g., using suitable integration algorithms). In addition, a rough approximation is available based on an Edgeworth expansion (see Harkness and Harkness [11])

$$F(x; \rho) \approx \Phi(x/\sqrt{\rho}) + \frac{1}{12\rho} \Phi^{(3)}(x/\sqrt{\rho}).$$

The moment-generating function immediately follows from (3.1),

$$\mathcal{M}(t) = \left(\frac{1}{\cos(t)} \right)^{\rho}, \qquad |t| < \pi/2.$$

Since its characteristic function in (3.1) is analytic, all moments exist are finite and given by $\mathbb{E}(X^k) = \mathbf{i}^k \mathscr{C}_{GHS}^{(k)}(0)$. In particular,

$$\mathbb{E}(X) = \mathbb{E}(X^3) = 0, \quad \mathbb{E}(X^2) = \rho = Var(X) \text{ and}$$

$$\mathbb{E}(X^4) = 3\rho^2 + 2\rho, \quad \text{i.e. } m_4 = 3 + \frac{2}{\rho}.$$

Consequently, the fourth standardized moment m_4 increases as ρ decreases and is bounded below by three which is achieved if ρ tends to infinity, and unbounded above.

2. Random number generation: Because the quantile function admits no simple expression, generation of GHS random numbers relies on acceptance-rejection methods as suggested by Devroye [6] for $\rho \geq 1$. In contrast, for $0 < \rho < 1$ we can apply a result of Harkness and Harkness [11] which states that a GHS-variable can be constructed from a couple of standard normal variables (Y_1, Y_2) with correlation $0 < \rho < 1$ as follows

$$X \equiv \ln \left| \frac{Y_1/Y_2 - \rho}{\sqrt{1 - \rho^2}} \right|. \tag{3.4}$$

For $\rho \geq 1$, the rejection algorithm of Devroye [6] basically uses an envelope constructed with the normal distribution in the main body and with the exponential distribution in the tails. For reason of brevity, we skip details and refer to Devroye [6], instead. The final algorithm works as follows (Fig. 3.2):

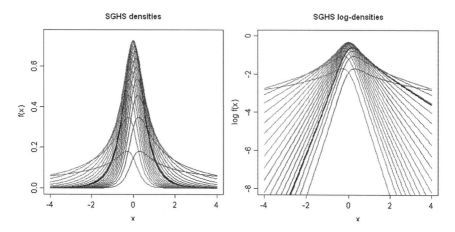

Fig. 3.2 SGHS distribution: Density and log-density for $t = 0.6$ and $\rho \in [-1.5, 1.5]$

Rejection method for the GHS distribution, $\rho \geq 1$
[Set-up.]
$t \leftarrow \rho^{5/8}$
$s \leftarrow \exp(1/(3\sqrt{\rho}))$
$C \leftarrow \sqrt{2\pi\rho}\,\frac{(\rho/e)^{\rho}}{\Gamma(\rho+1)}$
$p_n \leftarrow Cs$

$g_t \leftarrow g(t)$ where $g(x) := \frac{1}{\sqrt{2\pi\rho}}\left(1 + \frac{x^2}{\rho^2}\right)^{(\rho-1)/2}\exp\left(-x\arctan(x/\rho)\right)$

$\lambda \leftarrow g(t)/|g'(t)|$ (i.e., $\lambda \leftarrow (t/(\rho^2+t^2) + \arctan(t/\rho))^{-1}$)
$p_t \leftarrow 2Cg_t\lambda$
[Generator.]
repeat
      generate i.i.d. uniform [0,1] random variates U, V.
      if $U < \frac{p_n}{p_n+p_t}$
          then generate a standard normal random variate N
               set $X \leftarrow N\sqrt{\rho}$
               if $|X| > t$ then Accept $\leftarrow$ False
                    else $W \leftarrow Vp_n(2\pi\rho)^{-1/2}\exp\left(-X^2/2\rho\right)$
                        Accept $\leftarrow [W < Cg(X)]$
                        if not Accept
                              then Accept $\leftarrow [W < Csg(X)]$
                                    if Accept then
                                          Accept $\leftarrow [W < f(X)]$
          else generate an exponential random variate E
               set $X \leftarrow t + \lambda E$
               $W \leftarrow VCg_t\exp(-E)$
               Accept $\leftarrow [W < f(x)]$

if Accept, then with probability 1/2, set $X \leftarrow -X$

until Accept

return X

3.3 Introducing Skewness by Means of the Esscher Transformation

1. SGHS distributions: As the moment-generating function $\mathcal{M}(t)$ of a GHS-variable X exists, a very simple way to obtain a skew version is to apply the Esscher transformation to X in order to obtain the following *skew GHS distribution* (see, e.g., Grigelionis [8]) for $|h| < \pi/2$

$$
\begin{aligned}
f(x; \rho, h) &= \frac{e^{hx}}{\mathcal{M}_{GHS}(h)} \cdot f_{GHS}(x; \rho) \\
&= \frac{e^{hx}}{\cos(h)^{-\rho}} \cdot \frac{2^{\rho-2}}{\pi \, \Gamma(\rho)} \cdot \left| \Gamma\left(\frac{\rho}{2} + \mathbf{i}\frac{x}{2}\right) \right|^2 \\
&= \frac{2^{\rho-2}}{\pi \, \Gamma(\rho)} \cdot \left| \Gamma\left(\frac{\rho}{2} + \mathbf{i}\frac{x}{2}\right) \right|^2 \exp(hx + \rho \ln(\cos(h))). \quad (3.5)
\end{aligned}
$$

A more convenient parameterization results by setting $\beta \equiv \tan(h) \in (-\infty, \infty)$:

$$
f(x; \rho, \beta) = \left(\frac{1}{\sqrt{1 + \beta^2}}\right)^{\rho} \frac{2^{\rho-2}}{\pi \, \Gamma(\rho)} \cdot \left| \Gamma\left(\frac{\rho}{2} + \mathbf{i}\frac{x}{2}\right) \right|^2 \exp(\arctan(\beta)x), \quad (3.6)
$$

where we made use of the relation $\cos(\arctan(x)) = (\sqrt{1 + x^2})^{-1}$. This skew GHS distribution is known in the statistical literature as NEF-GHS or Laha-Lukacs distribution (e.g., Morris [20] or Jørgensen [14] pp. 100-103). Defining $\delta = \rho/2$ and starting from $\mathcal{M}(t) = (\cos(t/2))^{-2\delta}$ for $|t| < \pi$, an (equivalent) density given by

$$
f(x; \delta, h) = \frac{(2 \cos(h/2))^{2\delta}}{2\pi \, \Gamma(2\delta)} \cdot \exp(hx) \cdot |\Gamma(\delta + \mathbf{i}x)|^2 \quad (3.7)
$$

is known as the Meixner distribution in the mathematical and/or financial literature (e.g., Meixner [19] or Schoutens [23, 24]). Henceforth, (3.5) will used for further derivations.

2. Moment-generating functions, moments and tails: The moment-generating function of a SGHS variable exists and is derived as follows for $|h| < \pi/2$:

$$
\mathcal{M}(t) = \int_{-\infty}^{\infty} e^{tx} \frac{e^{hx}}{\cos(h)^{-\rho}} \cdot \frac{2^{\rho-2}}{\pi \, \Gamma(\rho)} \cdot \left| \Gamma\left(\frac{\rho}{2} + \mathbf{i}\frac{x}{2}\right) \right|^2 dx
$$

$$= \int_{-\infty}^{\infty} \frac{e^{(h+t)x}}{\cos(h)^{-\rho}} \cdot \frac{\cos(h+t)^{-\rho}}{\cos(h+t)^{-\rho}} \cdot \frac{2^{\rho-2}}{\pi \Gamma(\rho)} \cdot \left| \Gamma \left(\frac{\rho}{2} + i\frac{x}{2} \right) \right|^2 dx$$

$$= \frac{\cos(h+t)^{-\rho}}{\cos(h)^{-\rho}} \int_{-\infty}^{\infty} \frac{e^{(h+t)x}}{\cos(h+t)^{-\rho}} \cdot \frac{2^{\rho-2}}{\pi \Gamma(\rho)} \cdot \left| \Gamma \left(\frac{\rho}{2} + i\frac{x}{2} \right) \right|^2 dx$$

$$= \left(\frac{\cos(h)}{\cos(h+t)} \right)^{\rho} = \left(\frac{(\sqrt{1+\beta^2})^{-1}}{\cos(\arctan(\beta)+t)} \right)^{\rho}. \tag{3.8}$$

Consequently, all moments exist. In particular, the first four power moments are

$$\mathbb{E}(X) = \rho\beta,$$

$$\mathbb{E}(X^2) = \rho^2\beta^2 + \rho\beta^2 + \rho,$$

$$\mathbb{E}(X^3) = \rho^3\beta^3 + 3\rho^2\beta^3 + 3\rho^2\beta + 2\rho\beta^3 + 2\rho\beta,$$

$$\mathbb{E}(X^4) = \beta^2\rho^3(\rho\beta^2 + 6\beta^2 + 6) + \rho^2(11\beta^4 + 14\beta^2 + 3) + 2\rho(3\beta^4 + 4\beta^2 + 1)$$

From this, we conclude that $Var(X) = \rho(1 + \beta^2)$. Further, third and fourth standardized moments are given by

$$m_3 = \frac{\beta \left(\rho^2\beta^2 + 3\rho\beta^2 + 3\rho + 2\beta^2 + 2 \right)}{\left(1 + \beta^2 \right) \sqrt{\rho \left(1 + \beta^2 \right)}},$$

$$m_4 = \frac{\rho^3\beta^4 + 6\rho^2\beta^4 + 6\rho^2\beta^2 + 11\rho\beta^4 + 14\rho\beta^2 + 3\rho + 6\beta^4 + 8\beta^2 + 2}{\rho \left(1 + \beta^2 \right)^2}.$$

In terms of the second parameterization (see Grigelletto and Provasi [10]), skewness and kurtosis "simplify" to

$$m_3 = \sin(h) \sqrt{\frac{1}{\delta(\cos(h)+1)}} \quad \text{and} \quad m_4 = 3 - \frac{\cos(h) - 2}{\delta}.$$

Grigelionis [9] shows that SGSH distributions have semi-heavy tails, i.e.,

$$f(x; \rho, h) \sim \begin{cases} C_- |x|^{\rho-1} \exp(-(\pi - h)|x|) & \text{as } x \to -\infty, \\ C_- |x|^{\rho-1} \exp(-(\pi + h)|x|) & \text{as } x \to +\infty. \end{cases}$$

This SGHS family results as a special case from Grigelionis's [9]. Generalized z (GZ) distributions with characteristic function

$$\mathscr{C}_{GZ}(t) = \left(\frac{B(\beta_1 + \frac{it}{2\pi}, \beta_2 - \frac{it}{2\pi})}{B(\beta_1, \beta_2)} \right)^{\rho}$$

setting $\beta_1 = 0.5 + \frac{h}{2\pi}$ and $\beta_2 = 0.5 - \frac{h}{2\pi}$ (see also Mazzola and Muliere [17]). Moreover, for $\delta = 1$ the exponential generalized Beta distribution of the second kind (EGB2) is included, see McDonald [18] and Chap. 4.

3. Random number generation: Devroye [6] developes an algorithm based on the acceptance-rejection technique for $\rho \geq 1$ and $\lambda > 0$ which works as follows:

Generator for the NEF-GHS distribution, repeat
        generate U, V i.i.d. uniformly on [0,1]
        if $U < p_l/(p_l+p_m+p_r)$
            then generate E exponential
                $X \leftarrow t_l - E/\lambda_l$
                $T \leftarrow V g(t_l) \exp(-E)$
        else if $U > (p_l+p_m)/(p_l+p_m+p_r)$
            then generate E exponential
                $X \leftarrow t_r - E/\lambda_r$
                $T \leftarrow V g(t_r) \exp(-E)$
        else generate W uniformly on [0,1]
            $W \leftarrow t_l + 1/\lambda_m \ln(1 - W(1 - \exp(2\delta\lambda_m)))$
            (if $\lambda_m = 0$, set $X \leftarrow t_l + 2\delta W$)
            $T \leftarrow V g(t_m) \exp(\lambda_m(X - t_m))$
            (if $\lambda_m = 0$, set $T \leftarrow V p_m/2\delta$)
        Accept $\leftarrow [T < g(X) \exp(-1/3\rho)]$ ('quick accept')
        if not Accept
            then Accept $\leftarrow [T < g(X)]$ ('quick reject')
                if Accept then Accept $\leftarrow [T < f(x)]$
until Accept
return X

Alternatively, Grigoletto and Provasi [10] propose to approximate the Meixner density with Johnson's S_U translation system (see Johnson [12]) which is highly flexible and, through its functional forms, is able to closely approximate many heavy-tailed continuous distributions. Also, simulating from this distribution is relatively simple.

4. Characterizations: A first characterization of the SGHS distribution is connected to so-called natural exponential families, briefly NEF's (see, e.g., Barndorff-Nielsen [2] or Morris and Lock [21]). NEFs are parametric families of distributions with natural parameter $\theta \in \Theta$ where the corresponding random variable X satisfies

$$P_\theta(X \in A) = \int_A \exp(\theta x - \psi(\theta)) dF(x), \qquad (3.9)$$

for a certain function ψ and where F denotes a cumulative distribution function, without loss of generality. Restricting analysis to the NEF subclass with quadratic variance functions (briefly: NEF-QVF) of the form

$$Var(X) = V(\mu) = v_0 + v_1\mu + v_2\mu^2 \text{ with } \mu = \mathbb{E}(X), \qquad (3.10)$$

it is known (see Laha and Lukacs [15], Bolger and Harkness [4], Morris [20] and Slate [25]) that NEF-QVF has only six members, namely, normal, Poisson, binomial, Gamma, negative binomial, and SGHS distribution. For the latter, in (3.9) and (3.10)

$$\psi(x) = -\rho \ln(\cos(h)), \quad v_0 = 0, \ v_1 = \rho, \ v_2 = \rho^{-1}.$$

Note in addition, that $\psi'(\theta) = \mathbb{E}(X)$ and $\psi''(\theta) = Var(X)$. For a multivariate generalization of NEF-QVF's we refer to Casalis [5], whereas Letac and Mora [16] deal with NEF distributions with cubic variance function.

A second characterization originates in Meixner [19] from the theory of orthogonal polynomials: Morris [20] proves that $\{P_m\}_{m\geq0}$ with $P_0 = 1$ and

$$P_m(x, \mu) = V^m(\mu)\left\{\frac{d^m}{d\mu^m}f(x;\theta)\right\}/f(x;\theta), \quad m \geq 1 \qquad (3.11)$$

forms a family of orthogonal polynomials for a NEF-QVF density $f(x;\theta)$. For instance, assuming f in (3.11) to be Gaussian, Hermite polynomials appear. In case of SGHS densities f, the resulting polynomial turns out to be so-called Meixner-Pollaczek polynomials with

$$P_m^{\lambda}(x) \equiv \frac{(2\lambda)_m}{m!} \cdot e^{im\phi} \cdot {}_2F_1(-m, \lambda + \mathbf{i}x; 2\lambda; 1 - e^{-2i\phi})$$

and where ${}_2F_1$ denotes the hypergeometric function and $(\cdot)_m$ the Pochhammer symbol.

Meixner processes: Since SGHS distributions are infinitely divisible, a specific Lèvy process $\{X_t\}_t \geq 0$ can be associated. This was done by Schoutens and Teugels [22] and Grigelionis [8] and called as Meixner process. It starts at zero, i.e., $X_0 = 0$, has independent and stationary increments, i.e., $X_{t+s} - X_t = f(t)$, and X_t follows a SGHS or Meixner distribution with parameters $(\mu, \sigma, \beta, \delta t)$. Grigelionis [8] shows that this process has Lèvy triplet $(\gamma, 0, \nu)$ with

$$\gamma = \frac{\rho}{2}\tan(h/2) - \rho\int_1^\infty \frac{\sinh(hx)}{\sinh(\pi x)}dx, \quad \nu(dx) = \frac{\rho}{2}\left(\frac{\exp(hx)}{x\sinh(\pi x)}\right).$$

In particular, it has no Brownian part and a pure jump part governed by the Lèvy measure (see also Mazzola and Muliere [17]). Because $\int_{-1}^{1} |x| \nu(dx) = \infty$, the process is of infinite variation. Pricing of financial derivatives is one of the most popular applications of Meixner processes, see Schoutens [24].

References

1. Abramowitz, M., Stegun, I.A.: Handbook of Mathematical Functions with Formulas, Graphs, and Mathematical Tables. Courier Dover Publications, New york (1964)
2. Barndorff-Nielsen, O.E.: Information and Exponential Families in Statistical Theory. Wiley, New York (1978)
3. Baten, W.D.: The probability law for the sum of n independent variables, each subject to the law $(1/(2h))$ sech $(\pi x/(2h))$. Bull. Am. Math. Soc. **40**, 284–290 (1934)
4. Bolger, E.M., Harkness, W.L.: Characterizations of some distributions by conditional moments. Ann. Math. Stat. **36**, 703–705 (1965)
5. Casalis, M.: The $2d + 4$ simple quadratic natural exponential families on R. Ann. Stat. **24**, 1828–1854 (1996)
6. Devroye, L.: On random variate generation for the generalized hyperbolic secant distributions. Stat. Comput. **3**(3), 125–134 (1993)
7. Gradshteyn, I.S., Ryzhik, I.M.: Table of Integrals. Series and Products. Academic Press, San Diego (2000)
8. Grigelionis, B.: Processes of meixner type. Lith. Math. J. **39**(1), 33–41 (1999)
9. Grigelionis, B.: Generalized z-distributions and related stochastic processes. Lith. Math. J. **41**(3), 239–251 (2001)
10. Grigoletto, M., Provasi, C.: Simulation and estimation of the Meixner distribution. Commun. Stat. Simul. Comput. **38**(1), 58–77 (2009)
11. Harkness, W.L., Harkness, M.L.: Generalized hyperbolic secant distributions. J. Am. Stat. Assoc. **63**(321), 329–337 (1968)
12. Johnson, N.L.: Systems of frequency curves generated by methods of translation. Biometrika **36**, 146–176 (1949)
13. Johnson, N.L., Kotz, S.: Continuous Univariate Distributions, Vol. 2, Houghton-Mifflin, Boston (1970)
14. Jørgensen, B.: Theory of Dispersion Models. Chapman & Hall, London (1997)
15. Laha, R.G., Lukacs, E.: On a problem connected with quadratic regression. Biometrika **47**, 335–343 (1960)
16. Letac, G., Mora, M.: Natural real exponential families with cubic variance functions. Annals. Stat. **18**(1), 1–37 (1990)
17. Mazzola, E., Muliere, P.: Reviewing alternative characterizations of Meixner processes. Probab. Surv. **8**, 127–154 (2011)
18. McDonald, J.B.: Parametric models for partial adaptive estimation with skewed and leptokurtic residuals. Econometric Lett. **37**, 273–288 (1991)
19. Meixner, J.: Orthogonale Polygonsysteme mit einer besonderen Gestalt der erzeugenden Funktion. J. London Math. Soc. **9**, 6–13 (1934)
20. Morris, C.N.: Natural exponential families with quadratic variance functions. Ann. Stat. **10**(1), 65–80 (1982)
21. Morris, C.N., Lock, K.F.: Unifying the named natural exponential families and their relatives. Am. Stat. **63**(3), 247–253 (2009)

22. Schoutens, W., Teugels, J.L.: Lévy processes, polynomials and martingales. Commun. Stat. Stoch. Models. **14**(1&2), 335–349 (1998)
23. Schoutens, W.: The Meixner process in finance. EURANDOM-Report 2001–2002. http:// alexandria.tue.nl/repository/books/548458.pdf. (2001). Accessed 15 Jan 2012
24. Schoutens, W.: Lévy Processes in Finance - Pricing Financial Derivatives. Wiley Series in Probability and Statistics. Wiley, New York (2003)
25. Slate, E.H.: Parameterizations for Natural Exponential Families with Quadratic Variance Functions. J. Am. Stat. Assoc. **89**(428), 1471–1482 (1994)

Chapter 4
The BHS Distribution Family

Abstract The shape of a probability distribution is often characterized by the distribution's skewness and kurtosis. Starting from a symmetric "parent" density f on the real line, we can modify its shape (i.e., introduce skewness and in-/decrease kurtosis) if f is appropriately weighted. In particular, every density w on the interval $(0, 1)$ is a specific weighting function. In this chapter, we follow up a proposal of Jones [12] and choose the Beta distribution as underlying weighting function. "Parent" distributions like the Student-t, the logistic, and the normal distribution have already been investigated in the literature. Based on the assumption that f is the density of a hyperbolic secant distribution, we focus on the Beta-hyperbolic secant (BHS) distribution. In contrast to the Beta-normal distribution and to the Beta-Student-t distribution, BHS densities are always unimodal and all moments exist. In contrast to the Beta-logistic distribution, the BHS distribution is more flexible regarding the range of skewness and leptokurtosis combinations.

Keywords Definition and properties · Weighting function · Order statistics approach · EGB2 distribution

4.1 Introducing Skewness and Kurtosis via Order Statistics

Several techniques can be applied to symmetric distributions in order to generate asymmetric ones with possibly lighter or heavier tails. In terms of density functions—provided their existence—most of these methods can be represented by

$$g(x; \theta) = f(x)w(F(x); \theta), \tag{4.1}$$

where g denotes the transformed density, f and F the (symmetric) density and cumulative distribution function, respectively, of the original ("parent") distribution and w is an appropriate weighting function on the interval $(0, 1)$ with parameter vector θ

M. J. Fischer, *Generalized Hyperbolic Secant Distributions*,
SpringerBriefs in Statistics, DOI: 10.1007/978-3-642-45138-6_4,
© The Author(s) 2014

(see, for instance, Ferreira and Steel [8]). Choosing $w(u; \lambda) = 2F(\lambda F^{-1}(u))$, the skewing mechanism of Azzalini [2, 3] is recovered. Similarly, using

$$w(u; \lambda) = \frac{2}{\lambda + \frac{1}{\lambda}} \frac{f(\lambda^{\mathrm{sign}(0.5-u)} F^{-1}(u))}{f(F^{-1}(u))} \tag{4.2}$$

corresponds to applying different parameters of scale to the positive and the negative part of a symmetric density (see, for example, Fernández et al. [7] and Theodossiou [18]).

In particular, every probability density on $(0, 1)$ which is not uniform can be used either to introduce skewness and/or to modify the kurtosis of the parent distribution. A very attractive choice is the density of a Beta distribution, i.e.,

$$w(x; \beta_1, \beta_2) = \frac{1}{B(\beta_1, \beta_2)} x^{\beta_1 - 1} (1 - x)^{\beta_2 - 1}, \quad \beta_1, \beta_2 > 0, \tag{4.3}$$

where $B(a, b) = \int_0^1 t^{a-1}(1 - t)^{b-1} dt$ denotes the Beta function (cf. Jones [12]). Examples where (4.3) has been used in the literature include the following:

- Aroian [1], Prentice [17]: Beta-logistic distribution (which is also termed as exponential generalized beta of the second kind or EGB2 distribution, or $\ln F$ distribution), see also Sect. 4.4.
- Eugene et al. [6]: Beta-normal (BN) distribution,
- Jones and Faddy [13]: Beta-Student-t distribution.

Fischer and Vaughan [9] introduced the BHS distribution as a weighted hyperbolic secant distribution with weights from (4.3). Whereas both Beta-normal and Beta-Student-t distribution do not guarantee unimodality—except for a special parameterization given in Ferreira and Steel [8]—the BHS distribution does. In contrast to the Beta-Student-t distribution, all moments of the BHS distribution exist. Although the Beta-logistic and the BHS distribution are very similar, the BHS distribution is more flexible regarding skew and leptokurtic data, see Fischer and Vaughan [9].

Note that (4.3) can be replaced with

$$w(x, a, b) = abx^{a-1}(1 - x^a)^{b-1}, \quad a > 0, b > 0, \quad x \in [0, 1],$$

the density of the so-called the Kumaraswamy (K) distribution, see Kumaraswamy [15], which has properties similar to the beta distribution but has some advantages in terms of tractability. Whereas Cordeiro and de Castro [5] discuss, in particular, the K-normal distribution family, Fischer [10] focusses on the K-hyperbolic secant (briefly: KHS) case.

4.2 BHS Distribution: Definition

Recall from Chap. 1 that the probability density function of a hyperbolic secant distribution is given by

$$f_{HS}(x) = \frac{1}{\pi \cosh(x)} = \frac{2}{\pi(e^x + e^{-x})}, \quad x \in \mathbb{R}. \tag{4.4}$$

It is symmetric and the corresponding cumulative distribution function is

$$F_{HS}(x) = \frac{2 \arctan(e^x)}{\pi}. \tag{4.5}$$

The inverse cumulative distribution function is $F_{HS}^{-1}(u) = \ln(\tan(\frac{\pi u}{2}))$. Combining (4.1), (4.3), (4.4), and (4.5), the density of the Beta-hyperbolic secant (BHS) distribution is defined by

$$f(x; \beta_1, \beta_2) = \frac{B(\beta_1, \beta_2)^{-1}}{\pi \cosh(x)} \frac{\left[\frac{2}{\pi} \arctan(\exp(x))\right]^{\beta_1 - 1}}{\left[1 - \frac{2}{\pi} \arctan(\exp(x))\right]^{1-\beta_2}}, \tag{4.6}$$

where $\beta_1 > 0$ and $\beta_2 > 0$ determine the shape of the density. The corresponding cumulative distribution function is

$$F(x; \beta_1, \beta_2) = \frac{B_{F^{-1}(x)}(\beta_1, \beta_2)}{B(\beta_1, \beta_2)} \quad \text{with} \quad B_u(p, q) = \int_0^u t^{p-1}(1 - t)^{q-1} dt.$$

Introducing a location parameter $\mu \in \mathbb{R}$ and a scale parameter $\sigma > 0$, the BHS density from (4.6) generalizes to

$$f(x) = \frac{B(\beta_1, \beta_2)^{-1}}{\sigma \pi \cosh(\frac{x-\mu}{\sigma})} \left[\frac{2}{\pi} \arctan(e^{\frac{x-\mu}{\sigma}})\right]^{\beta_1 - 1} \left[1 - \frac{2}{\pi} \arctan(e^{\frac{x-\mu}{\sigma}})\right]^{\beta_2 - 1}.$$

Different densities and their corresponding log-densities with $\mu = 0$, $\sigma = 1$, $\beta_1 = 1$, and varying β_2 are plotted in Fig. 4.1.

The BHS distribution with parameters $\mu, \sigma, \beta_1, \beta_2$ is symmetric about μ for $\beta \equiv \beta_1 = \beta_2$. Moreover, it is skewed to the right for $\beta_1 > \beta_2$ and skewed to the left for $\beta_1 < \beta_2$. Assume that $\beta_1 = \beta_2 \equiv \beta$. Then, kurtosis increases if β decreases and vice versa. First of all, for $\beta_1 = \beta_2 = 1$ the hyperbolic secant distribution is recovered. Setting $\beta_2 = 1$ or $\beta_1 = 1$, skew hyperbolic secant distributions can be obtained. A generalized symmetric family of hyperbolic secant distributions is achieved for $\beta_1 = \beta_2 = \beta$, where β governs the amount of kurtosis. Like the Beta-logistic distribution and the Beta-normal distribution, the BHS distribution converges to the normal distribution for $\beta_1, \beta_2 \to \infty$.

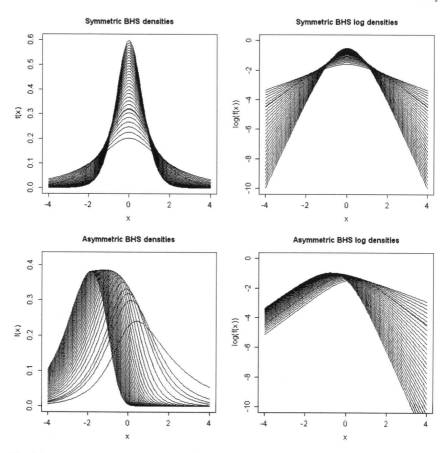

Fig. 4.1 BHS distribution: Different densities and log-densities $\beta_1 = \beta_2 \in [0.5, 3]$ (*upper panels*) and $\beta_1 = 1$, $\beta_2 \in [1, 8]$ (*lower panels*)

4.3 BHS Distribution: Properties

The BHS distribution has exponentially decaying tails. In particular, the log-density is asymptotically linear with slope determined by β_1 and β_2, respectively. In particular, it can be shown that for large x

$$f(x; \beta_1, \beta_2) \sim C \exp(-x) \exp((1 - \beta_2)x) = C \exp(-\beta_2 x), \quad C = \frac{(2/\pi)^{\beta_2}}{B(\beta_1, \beta_2)}.$$

Further, $\beta_2 < 1$ corresponds to distributions with heavier than plain exponential tails, while if $\beta_2 > 1$ the distributions have lighter than plain exponential tails. Obviously, the exponential tail behavior of the BHS distribution guarantees the existence of all moments. In particular, the mth non-central moment of a BHS density is given by

$$\mathbb{E}(X^m) = \frac{1}{B(\beta_1, \beta_2)} \int_0^1 \ln^m(\tan(\frac{\pi}{2}u))u^{\beta_1-1}(1-u)^{\beta_2-1}du.$$

From Gradshteyn and Ryhzik [11], formula 1.518.3 and 9.616 we can write

$$\tan(\frac{\pi}{2}u) = \ln(\frac{\pi}{2}u) + \sum_{k=1}^{\infty} \frac{(2^{2k-1}-1)\zeta(2k)}{k2^{2k-1}}u^{2k} = \ln(\frac{\pi}{2}u) + u^2 \sum_{k=0}^{\infty} a_k u^{2k}$$

with the usual Riemann zeta function

$$\zeta(2k) = \sum_{l=1}^{\infty} \frac{1}{l^{2k}} \text{ and } a_k = \frac{(2^{2k+1}-1)\zeta(2k+2)}{(k+1)2^{2k+1}}. \tag{4.7}$$

Using the notation

$$\frac{\partial^v}{\partial p^v}B(p,q) \equiv B^{v,0}(p,q), \quad B^{0,0}(p,q) = B(p,q),$$

Fischer and Vaughan [9] show that for $m > 0$

$$\mathbb{E}(X^m) = \frac{1}{B(\beta_1, \beta_2)} \left[\sum_{j=0}^{m} \binom{m}{j} \ln^{m-j}(\frac{\pi}{2}) B^{j,0}(\beta_1, \beta_2) \right.$$

$$\left. + \sum_{k=0}^{\infty} \sum_{j=1}^{m} \binom{m}{j} a_k^{(j)} \sum_{i=0}^{m-j} \binom{m-j}{i} \ln^{m-j-i}(\frac{\pi}{2}) B^{i,0}(2k+2j+\beta_1, \beta_2) \right],$$

where

$$a_0^{(j)} = a_0^j, \quad a_k^{(j)} = \frac{1}{ka_0} \sum_{i=1}^{k}(ij-k+i)a_i a_{k-i}^{(j)}, \quad k \geq 1.$$

Thus, the mean of the BHS distribution is given by

$$\mathbb{E}(X) = \ln(\frac{\pi}{2}) + \psi(\beta_1) - \psi(\beta_1+\beta_2) + \sum_{k=0}^{\infty} a_k \frac{B(2k+2+\beta_1, \beta_2)}{B(\beta_1, \beta_2)}. \tag{4.8}$$

with a_k from (4.7). Note that ψ denotes the digamma function in the last equation. In contrast to (4.8), the corresponding formula for the Beta-logistic distribution is given by

$$\mathbb{E}(X) = \psi(\beta_1) - \psi(\beta_2).$$

From the first four moments, we can deduce the skewness and kurtosis coefficients M_3 and M_4 (i.e., the third and fourth standardized moments) for different parameter combinations of the BHS distribution.

Additionally, the score function can be derived for the BHS distribution which plays an important role in the theory of rank tests (see, e.g., Kravchuk [14]) for $\beta_1 = \beta_2 = 1$) Specifically, with $\zeta(x) \equiv \arctan(e^x)$ the score function of a BHS variable is given by

$$
\begin{aligned}
\psi(x; \beta_1, \beta_2) &= -\frac{g'(x; \beta_1, \beta_2)}{g(x; \beta_1, \beta_2)} \\
&= \frac{\tanh(x)\,\zeta(x)(e^{2x}+1)(2\zeta(x) - \pi) + e^x\beta_1(\pi - 2\zeta(x))}{(1 + e^{2x})\,\zeta(x)\,(2\,\zeta(x) - \pi)} \\
&\quad - \frac{e^x\pi - 2e^x\zeta(x)(2 - \beta_2)}{(1 + e^{2x})\,\zeta(x)\,(2\,\zeta(x) - \pi)}.
\end{aligned}
$$

Setting $\beta_1 = \beta_2 = 1$, the last equation reduces to $\psi(x; 1, 1) = \tanh(x)$.

Fischer and Vaughan [9] showed that BHS densities are unimodal for all $\beta_1, \beta_2 > 0$. This is not valid for the Beta-normal and the Beta-Student-t distribution, in general.

4.4 EGB2 Distribution

As already mentioned in the introduction of this chapter, the exponential generalized beta of the second kind (EGB2) distribution or Beta-logistic distribution (see, e.g., Aroian [1] or Prentice [17]) resembles the BHS distribution. Its density is of the form

$$
f(x; \beta_1, \beta_2) = \frac{1}{B(\beta_1, \beta_2)} \frac{\exp(\beta_1 x)}{(1 + \exp(x))^{\beta_1 + \beta_2}}.
$$

Also, its moment-generating function and, hence, all moments exist. In particular, skewness and kurtosis coefficients admit simple forms,

$$
m_3 = \frac{\psi''(\beta_1) - \psi''(\beta_2)}{(\psi'(\beta_1) - \psi'(\beta_2))^{1.5}} \in [-2, 2], \quad m_4 = \frac{\psi'''(\beta_1) + \psi'''(\beta_2)}{(\psi'(\beta_1) - \psi'(\beta_2))^2} \in [3, 9],
$$

where $\psi'(x)$, $\psi''(x)$, and $\psi'''(x)$ denotes the trigamma, tetragamma, and pentagamma functions, respectively, where $\psi^{(n)}(x) = \frac{d^n}{dx^n} \ln(\Gamma(x))$. Although, this family is commonly associated with a generalization of the classical logistic distribution (which arises on setting $\beta_1 = \beta_2 = 1$), it can also be interpreted as a generalized HS distribution, because this family is recovered for $\beta_1 = \beta_2 = 0.5$, and with specific scale parameter $\sigma = 1/\sqrt{2\pi}$ (which is not included in the density representation above). For additional properties of the EGB2 distribution, we refer to Barndorff-Nielsen et al. [4] or McDonald [16].

References

1. Aroian, L.A.: A study of R. A. Fisher's z distribution and the related F distribution. Ann. Math. Stat. **12**, 429–448 (1941)
2. Azzalini, A.: A class of distributions which includes the normal ones. Scand. J. Stat **12**, 171–178 (1985)
3. Azzalini, A.: Further results on a class of distributions which includes the normal ones. Statistica **46**, 199–208 (1985)
4. Barndorff-Nielsen, O.E., Kent, J., Sœrensen, B.: Normal variance-mean mixtures and z distributions. Int. Stat. Rev. **50**, 145–159 (1982)
5. Cordeiro, G.M., de Castro, M.: A new family of generalized distributions. J. Stat. Comput. Simul. **81**, 883–898 (2011)
6. Eugene, N., Lee, C., Famoye, F.: Beta-normal distribution and its applications. Commun. Stat. Theor. Methods **31**(4), 497–512 (2002)
7. Fernández, C., Osiewalski, J., Steel, M.F.J.: Modelling and inference with v-spherical distributions. J. Am. Stat. Assoc. **90**, 1331–1340 (1995)
8. Fernández, C., Steel, M.F.J.: A constructive representation of univariate skewed distributions. J. Am. Stat. Assoc. **101**(474), 823–829 (2006)
9. Fischer, M., Vaughan, D.C.: The Beta-hyperbolic secant (BHS) distribution. Austrian J. Stat. **39**(2), 245–258 (2010)
10. Fischer, M.: The Kumaraswamy-hyperbolic secant (KHS) distribution, with application. Working Paper (unpublished) (2013)
11. Gradshteyn, I.S., Ryzhik, I.M.: Table of Integrals Series and Products. Academic Press, San Diego (2000)
12. Jones, M.C.: Families of distributions arising from distributions of order statistics. Test **13**(1), 1–43 (2004)
13. Jones, M.C., Faddy, M.J.: A skew extension of the t-distribution, with applications. J. Roy. Stat. Soc. Ser. B **65**(1), 159–174 (2003)
14. Kravchuk, O.Y.: Rank test of location optimal for hyperbolic secant distribution. Commun. Stat. (Theory and Methods) **34**(7), 1617–1630 (2003)
15. Kumaraswamy, P.: Generalized probability density-function for double-bounded random-processes. J. Hydrol. **462**, 79–88 (1980)
16. McDonald, J.B.: Parametric models for partial adaptive estimation with skewed and leptokurtic residuals. Econometric Lett. **37**, 273–288 (1991)
17. Prentice, R.L.: Discrimination among some parametric models. Biometrika **62**, 607–614 (1975)
18. Theodossiou, P.: Financial data and the skewed generalized t distribution. Math. Sci. **44**(12), 1650–1660 (1975)

Chapter 5
The SHS and SASHS Distribution Family

Abstract In the preceding chapters, the hyperbolic secant density was directly manipulated by certain weighting functions in order to allow for skewness and flexible kurtosis. In contrast, the random variable itself might be transformed with a suitable transformation T. With focus on the standard normal distribution, this idea dates back to Tukey [16] and Hoaglin [9], who postulated reasonable requirements on T. Corresponding examples are Hoaglin's $GH-$transformation or the $GK-$transformation of Rayner and MacGillivray [12], which can be integrated in a generalized parameterization (see Fischer [4, 5]). Unfortunately, the corresponding densities admit no closed-form representation, in general. To overcome this problem, one might use Johnson's [10] $S-$transformation or the SAS transformation of Jones and Pewsey [11] which guarantee that each of the probability density, cumulative distribution, and quantile functions has a simple form. In contrast to Rieck and Nedelman (2008) and Jones and Pewsey [11], who apply the $S-$transformation and $SAS-$transformation, respectively, to the classical Gaussian distribution, this chapter is dedicated to $S-$ and $SAS-$transformed hyperbolic secant distributions which are the subject of Fischer and Herrmann (2011) and Fischer [6].

Keywords Definition and properties · Variable transformation · Tukey · S–transformation

5.1 Variable Transformations Based on the Sinus Hyperbolic Function

Henceforth, the focus is on direct transformations T of a hyperbolic secant variable Z. In case of strictly monotone increasing transformations, the corresponding inverse transformations T^{-1} exist and the derivation of the characterizing functions of the new random variable $X \equiv T(Z)$ is straightforward, see box below.

M. J. Fischer, *Generalized Hyperbolic Secant Distributions*,
SpringerBriefs in Statistics, DOI: 10.1007/978-3-642-45138-6_5,
© The Author(s) 2014

Variable transformations: Let Z denote a symmetric variable. For a given monotone transformation $T : \mathbb{R} \to \mathbb{R}$, the cumulative distribution function of $X = T(Z)$ is given by

$$F_X(x) = P(X \leq x) = P(T^{-1}(X) \leq T^{-1}(x)) = P(Z \leq T^{-1}(x)) = F_Z(T^{-1}(x)).$$

From that, the corresponding density reads as

$$f_X(x) = f_Z(T^{-1}(x)) \left| \frac{dT^{-1}(x)}{dx} \right| \tag{5.1}$$

and the quantile function as $F_X^{-1}(x) = T(F_Z^{-1}(x))$.

Assume that Z follows a hyperbolic secant distribution. In order to achieve a closed form for X, transformations based on the hyperbolic sine function could be taken into consideration. Johnson [10] introduces the so-called inverse hyperbolic sine transformation (briefly S−transformation, henceforth) as

$$S(x) \equiv S_{\theta,\beta}(x) = \sinh(\theta^{-1}(x + \beta)), \quad \beta \in \mathbb{R}, \theta > 0. \tag{5.2}$$

It is strictly monotone increasing (see Fig. 5.1) because $S'_{\theta,\beta}(x) = \theta^{-1} \cosh(\theta^{-1}(x + \beta)) > 0$. Its inverse is given by $S_{\theta,\beta}^{-1}(x) = \theta \operatorname{asinh}(x) - \beta$, where

$$\operatorname{asinh}(x) = \sinh^{-1}(x) = \ln(x + \sqrt{x^2 + 1}) \text{ with } \operatorname{asinh}'(x) = \frac{1}{\sqrt{x^2 + 1}}.$$

Similarly, Jones and Pewsey [11] consider the sinh-arcsinh (SAS) transformation

$$SAS(x) \equiv SAS_{\theta,\beta}(x) = \sinh\left(\theta^{-1}(\operatorname{asinh}(x) + \beta)\right), \quad \beta \in \mathbb{R}, \theta > 0. \tag{5.3}$$

Obviously, $SAS_{\theta,\beta}(x) = S_{\theta,\beta}(S_{1,0}^{-1}(x))$. This transformation is also strictly monotone increasing (see Fig. 5.1), because $\frac{d}{dx} S_{\theta,\beta}^{-1}(x) = \frac{\theta}{\sqrt{x^2+1}} > 0$. Moreover, notice that

$$SAS'_{\theta,\beta}(x) = \frac{\cosh\left(\theta^{-1}(\operatorname{asinh}(Z) + \beta)\right)}{\theta \sqrt{x^2 + 1}} \geq 0, \tag{5.4}$$

$$SAS_{\theta,\beta}^{-1}(x) = \sinh\left(\theta \operatorname{asinh}(x) - \beta\right) \text{ and } \frac{d}{dx} SAS_{\theta,\beta}^{-1}(x) = \frac{\cosh\left(\theta \operatorname{asinh}(x) - \beta\right)\theta}{\sqrt{x^2 + 1}}.$$

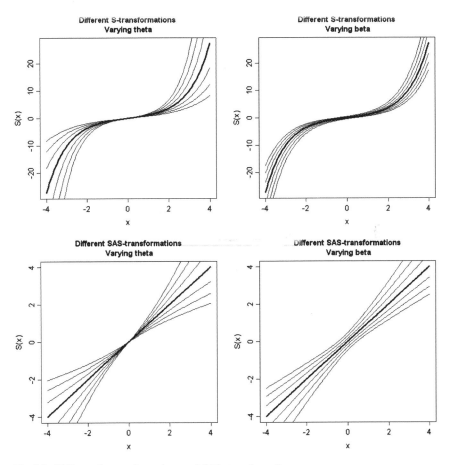

Fig. 5.1 Different $S-$transformations and $SAS-$transformations

5.2 Definition of the SHS and SASHS Distribution Family

SHS distribution: Originally, the S-transformation was applied to the normal distribution (see, for instance, Choi and Nam [2], Hansen et al. [8] or Rieck and Nedelman [13]) to model skew and heavy tailed data. Above that, Tadikamalla and Johnson [15] applied it to the logistic distribution and called it the L_U distribution. Combining the $S-$transformation with the hyperbolic secant variable according to (5.1), Fischer [6] discusses the SHS density (see Fig. 5.2) which takes the form

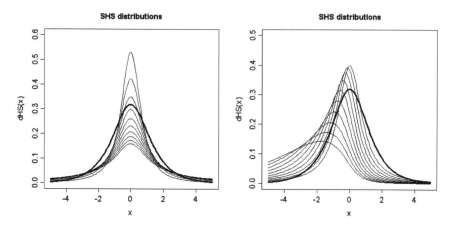

Fig. 5.2 Different SHS distributions with $\beta = 0, \theta \in [0.6, 2]$ (*left panel*) and $\theta = 0.8, \beta \in [0, 2]$ (*right panel*)

$$f(x) = \frac{1}{\pi \, \cosh(\theta \ln(x + \sqrt{x^2 + 1}) - \beta)} \frac{\theta}{\sqrt{x^2 + 1}}$$

$$= \frac{2/\pi}{\left((x + \sqrt{x^2 + 1})^\theta e^{-\beta} + (x + \sqrt{x^2 + 1})^{-\theta} e^\beta\right)} \frac{\theta}{\sqrt{x^2 + 1}}$$

with corresponding cumulative distribution function

$$F(x) = \frac{2}{\pi} \arctan \left(\exp(\theta \operatorname{asinh}(x) - \beta)\right)$$

$$= \frac{2}{\pi} \arctan \left(e^{-\beta} \left(x + \sqrt{x^2 + 1}\right)^\theta\right) \tag{5.5}$$

and quantile function, respectively,

$$F^{-1}(u) = \sinh \left(\frac{\ln(\tan(\pi u/2)) + \beta}{\theta}\right)$$

$$= \frac{1}{2} \left(e^{\beta/\theta} (\tan(\pi u/2))^{1/\theta} - e^{\beta/\theta} (\tan(\pi u/2))^{-1/\theta}\right). \tag{5.6}$$

Whereas β determines the skewness of the SHS distribution, $\theta > 0$ influences both tail thickness and peakedness.

SASHS distribution: Jones and Pewsey [11] apply the *SAS*-transformation to introduce skewness and kurtosis into a normally distributed random variable and show that the parameters β and θ in the distribution of the transformed random variable X act, respectively, as skewness and kurtosis parameters in the sense of corresponding orderings defined in van Zwet [17]. Similarly, Rosco et al. [14] apply

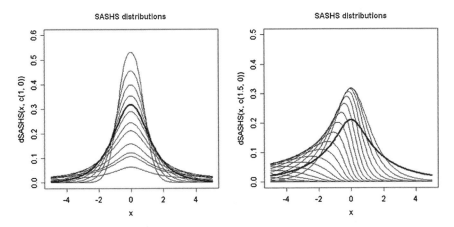

Fig. 5.3 Different SASHS distributions with $\beta = 0$, $\theta \in [0.6, 5]$ (*left panel*) and $\theta = 1$, $\beta \in [0, 20]$ (*right panel*)

the restricted *SAS*-transformation with $\theta = 1$ but arbitrary β to skew a Student-t variable with $v > 0$ degrees of freedom. Alternatively, Fischer and Herrmann [7] focus on the HS distribution. The corresponding SASHS density (see Fig. 5.3) reads as:

$$f(x) = \frac{\cosh{(\theta \, \mathrm{asinh}(x) - \beta)}\, \theta}{\pi \, \cosh(\sinh{(\theta \, \mathrm{asinh}(x) - \beta))}} \frac{1}{\sqrt{x^2 + 1}}. \tag{5.7}$$

Notice that

$$\cosh(\theta \, \mathrm{asinh}(x) - \beta) = \frac{1}{2}\left(e^{\theta \ln(x+\sqrt{x^2+1})-\beta} + e^{-\theta \ln(x+\sqrt{x^2+1})+\beta}\right)$$

$$= \frac{1}{2}\left(e^{-\beta}\left(x + \sqrt{x^2+1}\right)^{\theta} + e^{\beta}\left(x + \sqrt{x^2+1}\right)^{-\theta}\right).$$

Hence, the cumulative distribution function of a SASHS distribution is given by:

$$F(x) = \frac{2}{\pi} \arctan{(\exp(\sinh(\theta \, \mathrm{asinh}(x) - \beta)))}$$

$$= \frac{2}{\pi} \arctan\left[\exp\left(\frac{1}{2}\left\{e^{-\beta}\left(x + \sqrt{x^2+1}\right)^{\theta} - e^{\beta}\left(x + \sqrt{x^2+1}\right)^{-\theta}\right\}\right)\right]$$

and, hence, its inverse is given by:

$$F^{-1}(u) = \sinh\left(\frac{\mathrm{asinh}(\ln(\tan(\pi u/2))) + \beta}{\theta}\right)$$

$$= \frac{1}{2}\left(e^{\beta/\theta}\,[\ln{(\tan(\pi u/2))}]^{1/\theta} - e^{-\beta/\theta}\,[\ln(\tan(\pi u/2))]^{-1/\theta}\right).$$

5.3 Basic Properties of the SHS and SASHS Distribution Families

SHS distribution: First of all, the median depends only on β and θ and is given by:

$$x_{0.5} = \frac{\beta}{\theta}. \tag{5.8}$$

It can be shown that all SHS densities are unimodal (see Fischer [6]). Tails of this family of distributions are of the form

$$f(x) \overset{x \to \pm\infty}{\sim} \frac{C_0}{C_1 x^{1+\theta} + C_2 x^{1-\theta}}$$

for suitable constants C_0, C_1, and C_2. Hence, tails decrease as a power of $|x|$ such that the SHS family behaves in a way similar to Student's t distribution and moments exist only up to a certain order (depending on θ, see below). Provided their existence, the moments of the SHS family can be derived as follows: For $n \in \mathbb{N}$ notice that

$$
\begin{aligned}
S(z)^n &= \frac{1}{2^n} \left(e^{\theta^{-1}(z+\beta)} - e^{-\theta^{-1}(z+\beta)} \right)^n \\
&= \frac{1}{2^n} \sum_{i=0}^{n} \binom{n}{i} \left(e^{\theta^{-1}(z+\beta)} \right)^{n-i} \left(e^{-\theta^{-1}(z+\beta)} \right)^{i} (-1)^i \\
&= \frac{1}{2^n} \sum_{i=0}^{n} \binom{n}{i} \left(e^{\theta^{-1}(z+\beta)(n-i)} \right) \left(e^{-\theta^{-1}(z+\beta)i} \right) (-1)^i \\
&= \frac{1}{2^n} \sum_{i=0}^{n} \binom{n}{i} \left(e^{\theta^{-1}(z+\beta)(n-i)-\theta^{-1}(z+\beta)i} \right) (-1)^i \\
&= \frac{1}{2^n} \sum_{i=0}^{n} \binom{n}{i} \left(e^{\theta^{-1}(z+\beta)(n-2i)} \right) (-1)^i \\
&= \frac{1}{2^n} \sum_{i=0}^{n} \binom{n}{i} \left(e^{\frac{n-2i}{\theta}(z+\beta)} \right) (-1)^i = \frac{1}{2^n} \sum_{i=0}^{n} \binom{n}{i} \left(e^{\frac{n-2i}{\theta}\beta} \right) \left(e^{\frac{n-2i}{\theta}z} \right) (-1)^i.
\end{aligned}
$$

Replacing z by Z and taking expectations, we obtain

$$\mathbb{E}(X^n) = \mathbb{E}(S(Z)^n) = \frac{1}{2^n} \sum_{i=0}^{n} \binom{n}{i} \left(e^{\frac{n-2i}{\theta}\beta} \right) (-1)^i \mathscr{M}_Z \left(\frac{n-2i}{\theta} \right),$$

where

$$\mathscr{M}_Z(t) = \frac{1}{\cos(\pi t/2)} \quad \text{for } |t| < 1$$

denotes the moment-generating function of a hyperbolic secant variable. Provided their existence (i.e., for $n < \theta$), the first four power moments are given by

$$\mathbb{E}(X) = \frac{\sinh(\beta/\theta)}{\cos(\pi\theta/2)}, \quad \mathbb{E}(X^2) = \frac{1}{2}\left(\frac{\sinh(2\beta/\theta)}{\cos(\pi\theta)} - 1\right),$$

$$\mathbb{E}(X^3) = \frac{1}{4}\left(\frac{\sinh(3\beta/\theta)}{\cos(1.5\pi/\theta)} - 3\frac{\sinh(\beta/\theta)}{\cos(0.5\pi/\theta)}\right) \text{ and}$$

$$\mathbb{E}(X^4) = \frac{1}{8}\left(\frac{\cosh(4\beta/\theta)}{\cos(2\pi/\theta)} - 4\frac{\cosh(2\beta/\theta)}{\cos(\pi/\theta)} + 3\right).$$

From this, skewness and kurtosis (measured by third and fourth standardized moments) can be calculated in a straightforward manner.

SASHS distribution: In this case, the median admits the form

$$x_{0.5} = \sinh\left(\frac{\beta}{\theta}\right). \tag{5.9}$$

Asymptotically, the SASHS density behaves like

$$f(x) \overset{x\to\pm\infty}{\sim} K_0 \exp(-K_1 x^\theta) x^{\theta-1}$$

implying that this family has exponentially decaying tails, lighter than tails decreasing as a power of $|x|$ such that semi-heavy tail behaviour can be modeled. At first sight, SASHS families resemble the well-known generalized error distribution (GED) from Box-Tiao [1] or some related skew version, see, for instance, DiCiccio and Monti [3]. All moments exist and derive as follows: First, the n-th power of the SAS transformation can be written as

$$SAS(z)^n = \frac{1}{2^n} \sum_{i=0}^{n} \binom{n}{i} e^{\beta(n-2i)} \left(z + \sqrt{z^2+1}\right)^{\theta(n-2i)}$$

$$= \frac{1}{2^n} \sum_{i=0}^{n} \binom{n}{i} e^{\beta(n-2i)} \exp\left\{\theta(n-2i)\ln\left(\left(z+\sqrt{z^2+1}\right)\right)\right\}.$$

Replacing z by Z and taking expectations, we obtain

$$\mathbb{E}(X^n) = \frac{1}{2^n} \sum_{i=0}^{n} \binom{n}{i} e^{\beta(n-2i)} \mathbb{E}\left(\exp\left(\theta(n-2i)W\right)\right)$$

$$= \frac{1}{2^n} \sum_{i=0}^{n} \binom{n}{i} e^{\beta(n-2i)} \mathcal{M}_W(\theta(n-2i))$$

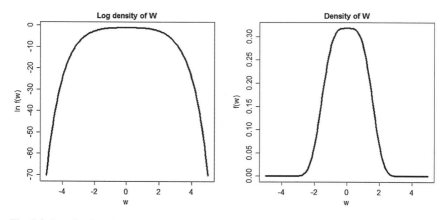

Fig. 5.4 Log density (*left panel*) and density (*right panel*) of W

with $W \equiv \operatorname{asinh}(X) = \ln(X + \sqrt{X^2 + 1})$ and where $\mathscr{M}_W$ denotes the corresponding moment-generating function. Some properties of W are summarized in the next box.

Consider the random variable $W \equiv T(X) \equiv \operatorname{asinh}(X) = \ln(X + \sqrt{X^2 + 1})$, where X represents the hyperbolic secant density. The underlying transformation T is monotone increasing because

$$T'(x) = \frac{1}{\sqrt{x^2 + 1}} > 0.$$

Because of its inverse function, $T^{-1}(x) = \sinh(x)$ with corresponding first derivative $(T^{-1})'(x) = \cosh(x) > 0$, the random variable W admits the following density (see Fig. 5.4)

$$f_W(w) = \frac{\cosh(w)}{\pi \cosh(\sinh(w))}, \quad w \in \mathbb{R}$$

with cumulative distribution function

$$F_W(w) = \frac{2}{\pi} \arctan(\exp(\sinh(w))), \quad w \in \mathbb{R}.$$

It has light tails (such that all moments exist). The first few even moments can be numerically approximated as:

$$\mathbb{E}(W) = 0, \quad \mathbb{E}(W^2) = 1.0655, \quad \mathbb{E}(W^3) = 0, \quad \mathbb{E}(W^4) = 2.587$$

$$\mathbb{E}(W^6) = 8.813, \quad \mathbb{E}(W^8) = 37.009.$$

References

1. Box, G.E.P., Tiao, G.C.: Bayesian Inference in Statistical Analysis. Addison-Wesley, Reading, Massachusetts (1973)
2. Choi, P., Nam, K.: Asymmetric and leptokurtic distribution for heteroskedastic asset returns: the S_U-normal distribution. J. Empirical Finance 15(1), 41–63 (2008)
3. DiCiccio, T.J., Monti, A.C.: Inferential aspects of the skew exponential power distribution. J. Am. Stat. Assoc. 99, 439–450 (2004)
4. Fischer, M.: A note on the construction of Tukey-type distributions. J. Stat. Res. 42(2), 79–88 (2008)
5. Fischer, M.: Generalized Tukey-type distribution with applications to financial return data. Stat. Pap. 1, 41–56 (2010)
6. Fischer, M.: A Skew and Leptokurtic Distribution with Polynomial Tails and Characterizing Functions in Closed Form. Working Paper (unpublished). Department of Statistics and Econometrics, FAU Erlangen-Nuremberg, Nuremberg (2012)
7. Fischer, M., Herrmann, K.: Two new skewed and leptokurtic distributions with explicit quantile functions for financial return data. Austrian J. Stat. (To appear) (2013)
8. Hansen, C.B., McDonald, J.B., Theodossiou, P.: Some flexible parametric models for partially adaptive estimators of econometric models. Econ. E-J. 7 (2007)
9. Hoaglin, D.C.: Summarizing shape numerically: the g and h distributions. In: Hoaglin, D.C., Mosteller, F., Tukey, J.W. (eds.) Data Analysis for Tables, Trends, and Shapes, pp. 461–513. Wiley, New York (1983)
10. Johnson, N.L.: Systems of frequency curves generazed by methods of translation. Biometrika 36, 149–176 (1949)
11. Jones, M.C., Pewsey, A.: Sinh-arcsinh distributions. Biometrika 96(4), 761–780 (2009)
12. Rayner, G.D., MacGillivray, H.L.: Numerical maximum likelihood estimation for the g-and-k and generalised g-and-h distributions. Stat. Comput. 12, 57–75 (2002)
13. Rieck, J.R., Nedelman, J.R.: A log-linear model for the Birnbaum-Saunders distributions. Technometrics 33, 51–60 (1991)
14. Rosco, J.F., Jones, M.C., Pewsey, A.: Skew t distributions via the sinh-arcsinh transformation. Test (2010). doi:10.1007/s11749-010-0222-2
15. Tadikamalla, P.R., Johnson, N.L.: Systems of frequency curves generated by transformations of logistic variables. Biometrika 69, 461–465 (1982)
16. Tukey, J.W.: The Practical Relationship Between the Common Transformations of Counts of Amounts. Technical Report No. 36. Princeton University Statistical Techniques Research, Princeton (1960)
17. van Zwet, W.R.: Convex Transformations of Random Variables. Mathematical Centre Tracts No. 7, Mathematical Centre, Amsterdam (1964)

Chapter 6
Application to Finance

Abstract Although able to incorporate different combinations of skewness and kurtosis, generalized hyperbolic secant distributions have mainly been neglected so far in the broad financial literature. Within this chapter we illustrate their flexibility in the context of different financial return distributions (e.g., stock indices and exchange rates). In particular, we compare its fit with that of popular benchmark models such as stable distributions, (skew) Student-t distributions and generalized hyperbolic distributions.

Keywords Financial return · GARCH model · Goodness-of-fit · Moment ratio plot

6.1 Excursion: Moment-Ratio Plots

Moment-ratio diagrams (MRD) were introduced for Pearson-type distributions by Elderton and Johnson [8] in order to provide a useful visual assessment of skewness and kurtosis. The classical moment ratio plot consists of all possible pairs of third and fourth standardized moments (M_3, M_4) that can be obtained through different combinations of the shape parameters of the underlying distributions, provided their existence. In general, the relation $M_3 < \sqrt{M_4 - 1}$ for $M_4 \geq 1$ holds, i.e., for a given level of kurtosis only a finite range of skewness may be spanned. For example, Fig. 6.1 contains MRD's for six generalized secant distributions, where all moments exists: SGSH1 and SGSH2 distribution (Chap. 2), SHS distribution (Chap. 3), Beta hyperblic secant (BHS) distribution and EGB2 distribution (Chap. 4), and SHS and SASHS distribution (Chap. 5). Except for EGB2 and BHS distribution where kurtosis (and hence skewness) is limited to 6 and 8, respectively, the other four candidates cover a broad area of the admissible range. In addition, SGSH1 and SGSH2 also admit platykurtic behavior. If third and fourth standardized moments are not defined, skewness-kurtosis plots based on alternative measures of skewness and tail weight may help, see Brys et al. [5, 6].

M. J. Fischer, *Generalized Hyperbolic Secant Distributions*,
SpringerBriefs in Statistics, DOI: 10.1007/978-3-642-45138-6_6,
© The Author(s) 2014

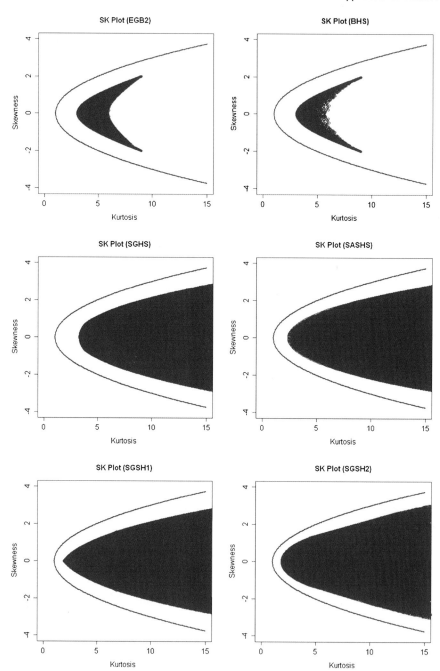

Fig. 6.1   Selected SK plots

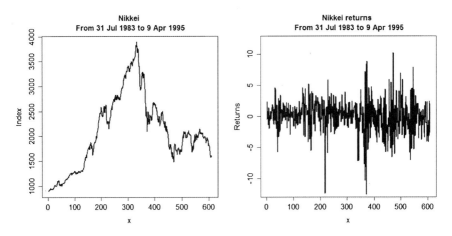

Fig. 6.2 Nikkei data: level versus returns

6.2 Return Series Under Consideration

1. Nikkei data: In order to adopt and compare estimation results for a great deal of distributions—in particular the stable distributions (STABLE)—priority is given to the weekly returns of the Nikkei from July 31, 1983 to April 9, 1995, with $N = 608$ observations. This series was intensively investigated, for example, by Mittnik et al. [20] because it exhibits typical stylized facts of financial return data. Figure 6.2 illustrates the time series of levels and corresponding log-returns.

2. FX data: Secondly, we chose data from foreign exchange markets (FX-markets) which are available from the PACIFIC Exchange Rate Service.[1] This service offered by Prof. Werner Antweiler at UBC's Sauder School of Business provides access to current and historic daily exchange rates through an online database retrieval and plotting system. In contrast to the volume notation, where values are expressed in units of the target currency per unit of the base currency,[2] the so-called *price notation* is used within this work which corresponds to the numerical inverse of the volume notation. All values are expressed in units of the base currency per unit of the target currency. Many European countries quote exchange rates this way.

Daily exchange rates for the EUR-USD are available from Jan 1, 2002 to Apr 30, 2012 ($n = 2593$). Figure 6.3 illustrates the corresponding time series for both levels and returns. For our analysis, we also consider two subperiods of this time series. First, from Jan 1, 2002 to Mar 31, 2008 ($n = 1568$) where the exchange rate steadily increases. Secondly, from Apr 1, 2008 to Apr 30, 2012 ($n = 1025$) where we observe a volatile period of down-movement. Table 6.1 summarizes the relevant statistics and Fig. 6.4 exhibits both histogram versus kernel density for the underlying series.

---

[1] Download under the URL-link http://fx.sauder.ubc.ca/.

[2] This is commonly used in North America to quote exchange rates.

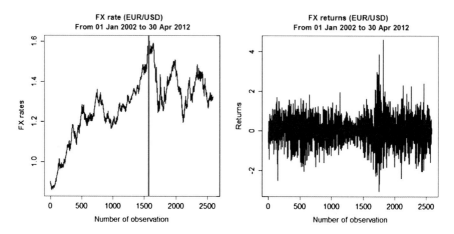

Fig. 6.3 EUR/USD exchange rate: level versus returns

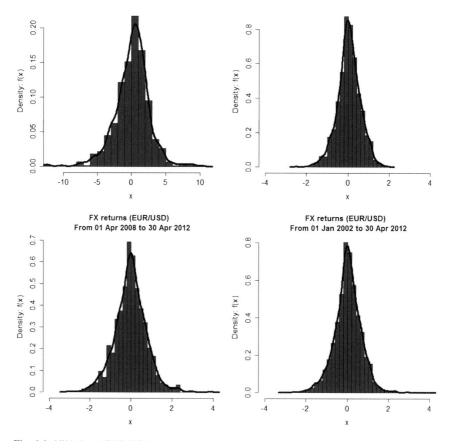

Fig. 6.4 Nikkei and EUR/USD returns: histogram versus kernel density

Table 6.1 Data set: descriptive and inductive data statistics

Data	No.	$\overline{X}$	S^2	$\mathbb{S}$	$\mathbb{K}$	$\mathscr{LB}$	$\mathscr{LM}$
Nikkei	608	0.0958	6.4365	−0.4517	6.1401	0.1675	0.0000
NikkeiGARCH	606	0.0340	0.8781	−0.6321	5.5857	0.1578	0.6635
EURUSD	2593	0.0147	0.4315	0.1012	5.4191	0.8940	0.0000
EURGARCH	2591	0.0343	0.9857	−0.0116	3.8534	0.9174	0.1984
EURUSD1	1568	0.0349	0.3100	−0.1738	3.7323	0.4819	0.0043
EURUSDGARCH1	1566	0.0715	0.9597	−0.0958	3.3494	0.7050	0.1356
EURUSD2	1025	−0.0162	0.6163	0.3035	5.3209	0.5459	0.0000
EURUSDGARCH2	1023	−0.0225	0.9975	0.1274	4.2596	0.9472	0.8033

6.3 Fitting Generalized Hyperbolic Secant Distribution: Unconditional Case

1. **Distributions under consideration**: The main purpose of this chapter is to compare the flexibility of the seven generalized hyperbolic secant distributions which where previously discussed, namely SGSH1 and SGSH2 (see Chap. 2 or Fischer [10, 11]), SGHS (see Chap. 2 or Morris [21]), BHS and EGB2 (see Chap. 4 or Fischer and Vaughan [12] and McDonald [19]), SASHS and SHS (see Chap. 5 or Fischer and Herrmann [14] and Fischer [13]). In addition, results are also provided for distribution families which have become popular in finance in the past: First, the Student-t distribution (T) and skew generalizations (ST, see Zhu and Galbraith [24]) where moments exist only up to a certain order. Secondly, stable distributions which appear from the central limit theorem if the variance does not exist (see, e.g., Nolan [22]). Finally the generalized hyperbolic (GH) distributions which were discussed by Prause [23] and include, for example, the Normal-inverse Gaussian distributions (see Barndorff-Nielsen [1, 2]) as well as the hyperbolic distributions (see Eberlein and Keller [7]) as special cases.

2. **Model setup and parameter estimation**: For the moment, assume that the underlying log-returns are independent and identically distributed, i.e.,

$$R_t = \mu + U_t \text{ with } U_t \sim D(0, \sigma^2, \eta), \quad t = 1, \dots, T,$$

where D denotes the underlying distribution model with location $\mu \in \mathbb{R}$, (constant) scale $\sigma > 0$ and shape parameter vector η. Define the vector of unknown parameters as $\Theta = (\mu, \sigma, \eta)$ and suppose that N observations $r_1, \dots, r_N$ are given. The corresponding log-likelihood function is defined as

$$LL(\theta) = \sum_{i=1}^{N} \ln (f_D(r_1, \dots, r_N; \Theta)).$$

Then, the maximum likelihood estimator (MLE) of Θ, indicated by $\widehat{\ell}_{ML}$ is the solution of the following optimization problem:

$$\widehat{\ell}_{ML} = \operatorname{argmax}_{\Theta} LL(\Theta).$$

Under certain regularity conditions it is known that (see Lehman and Casella [18]), if Θ_0 is the true and unknown parameter vector, the ML estimator $\widehat{\ell}_{ML}$ satisfies

$$\sqrt{N}(\widehat{\ell}_{ML} - \Theta_0) \xrightarrow{d} N(0, I_N^{-1}),$$

where I_N^{-1} denote the inverse of the so-called Fisher information matrix. A detailed treatment of estimating the Fisher information matrix can be found in Greene [17], Chap. 17. The empirical results of this note have been obtained with the statistical software package R using the constrained optimization function `nlminb` (see Gay [16]). Exemplarily, Appendix A includes the corresponding R code for the BHS distribution.

3. **Measuring goodness-of-fit**: Similar to Mittnik et al. [20], four criteria are employed to compare the goodness-of-fit of the different candidate distributions. The first is the *log-Likelihood value* (ℓ_N) obtained from the Maximum-Likelihood estimation. The ℓ_N-value can be considered as an "overall measure of goodness-of-fit and allows us to judge which candidate is more likely to have generated the data." As distributions with different numbers of parameters k are used, this is taken into account by calculating the *Akaike criterion* given by

$$AIC = -2 \cdot \ell_N + \frac{2N(k+1)}{N-k-2}.$$

The third criterion is the *Kolmogorov-Smirnov distance* as a measure of the distance between the estimated parametric cumulative distribution function, $\hat{F}$, and the empirical sample distribution, F_{emp}. It is usually defined by

$$\mathscr{K} = 100 \cdot \sup_{x \in \mathbb{R}} |F_{emp}(x) - \hat{F}(x)|. \tag{6.1}$$

Finally, the *Anderson-Darling statistic* is calculated, which weights $|F_{emp}(x) - \hat{F}(x)|$ by the reciprocal of the standard deviation of F_{emp}, namely $\sqrt{\hat{F}(x)(1 - \hat{F}(x))}$, that is

$$\mathscr{A}\mathscr{D}_0 = \sup_{x \in \mathbb{R}} \frac{|F_{emp}(x) - \hat{F}(x)|}{\sqrt{\hat{F}(x)(1 - \hat{F}(x))}}. \tag{6.2}$$

Instead of just the maximum discrepancy, the second and third largest value, which is commonly termed as $\mathscr{A}\mathscr{D}_1$ and $\mathscr{A}\mathscr{D}_2$, are also taken into consideration. Whereas $\mathscr{K}$ emphasizes deviations around the median of the fitted distribution, $\mathscr{A}\mathscr{D}_0$, $\mathscr{A}\mathscr{D}_1$ and $\mathscr{A}\mathscr{D}_2$ allow discrepancies in the tails of the distribution to be appropriately weighted.

Table 6.2 Goodness-of-fit for the unconditional and conditional case: Nikkei225

Distr.	k	ℓ_N	AIC	$\mathcal{K}$	$\mathcal{AP}_0$	$\mathcal{AP}_1$	$\mathcal{AP}_2$	ℓ_N	AIC	$\mathcal{K}$	$\mathcal{AP}_0$	$\mathcal{AP}_1$	$\mathcal{AP}_2$
NV	2	−1428.26	2862.56	6.889	4.971	2.839	1.081	−819.98	1646.00	6.673	21.943	1.002	0.210
HS	2	−1393.40	2792.84	4.310	0.216	0.150	0.121	−798.10	1602.25	3.243	0.186	0.100	0.098
GHS	3	−1392.25	2792.56	4.147	0.134	0.117	0.114	−798.10	1604.27	3.243	0.187	0.100	0.098
SGHS	4	−1388.08	2786.27	2.420	0.072	0.069	0.068	−791.60	1593.30	2.106	0.083	0.046	0.045
GSH	3	−1392.34	2792.74	4.178	0.142	0.117	0.115	−798.10	1604.27	3.241	0.192	0.099	0.099
SGSH1	4	−1387.54	2785.18	2.179	0.071	0.065	0.062	−795.11	1600.32	2.658	0.161	0.082	0.082
SGSH2	4	−1388.09	2786.28	2.412	0.073	0.071	0.069	−791.60	1593.31	2.129	0.082	0.046	0.045
BHS	4	−1387.99	2786.08	2.384	0.086	0.071	0.068	−791.52	1593.13	2.177	0.081	0.047	0.045
EGB2	4	−1388.11	2786.33	3.082	0.177	0.112	0.089	−791.60	1593.31	2.268	0.049	0.047	0.047
SHS	4	−1389.44	2788.98	2.842	0.083	0.080	0.078	−793.23	1596.55	2.248	0.062	0.061	0.061
SASHS	4	−1387.85	2785.78	2.368	0.070	0.068	0.066	−791.37	1592.84	2.035	0.085	0.046	0.044
GH	5	−1388.04	2788.22	2.434	0.072	0.070	0.069	−791.36	1594.85	2.124	0.108	0.050	0.046
T	2	−1392.25	2792.57	3.771	0.107	0.105	0.103	−797.86	1603.79	3.067	0.100	0.099	0.097
ST	4	−1387.90	2785.89	2.281	0.059	0.056	0.053	−791.72	1593.54	2.255	0.046	0.046	0.043
STAB	4	−1393.24	2796.57	3.284	0.074	0.072	0.068	−797.20	1604.50	3.303	0.067	0.064	0.064

Table 6.3 Goodness-of-fit for the unconditional and conditional case: FX EUR/USD (2002–2012)

Distr.	k	ℓ_N	AIC	$\mathscr{K}$	$\mathscr{AD}_0$	$\mathscr{AD}_1$	$\mathscr{AD}_2$	ℓ_N	AIC	$\mathscr{K}$	$\mathscr{AD}_0$	$\mathscr{AD}_1$	$\mathscr{AD}_2$
NV	2	−2589.07	5184.15	4.895	7.846	0.409	0.409	−3657.35	7320.71	3.360	0.083	0.074	0.072
HS	2	−2509.97	5025.95	1.164	0.056	0.054	0.053	−3653.87	7313.74	1.996	0.080	0.079	0.078
GHS	3	−2509.86	5027.74	1.291	0.054	0.052	0.051	−3640.09	7288.19	1.821	0.047	0.045	0.044
SGHS	4	−2509.50	5029.03	1.320	0.051	0.046	0.044	−3639.76	7289.54	1.746	0.040	0.038	0.038
GSH	3	−2509.87	5027.75	1.292	0.054	0.052	0.051	−3642.66	7293.33	1.394	0.059	0.057	0.057
SGSH1	4	−2509.65	5029.32	1.339	0.048	0.048	0.046	−3641.63	7293.27	1.740	0.046	0.046	0.046
SGSH2	4	−2509.51	5029.04	1.322	0.050	0.046	0.044	−3641.32	7292.66	1.707	0.041	0.041	0.041
BHS	4	−2509.58	5029.18	1.293	0.048	0.046	0.044	−3640.12	7290.25	1.829	0.040	0.039	0.038
EGB2	4	−2509.60	5029.22	1.262	0.047	0.045	0.044	−3640.24	7290.50	1.857	0.040	0.040	0.039
SHS	4	−2510.44	5030.90	1.230	0.052	0.050	0.049	−3653.82	7317.66	2.036	0.076	0.075	0.074
SASHS	4	−2509.68	5029.38	1.261	0.048	0.047	0.046	−3641.38	7292.77	1.699	0.045	0.044	0.043
GH	5	−2508.52	5029.07	1.150	0.055	0.054	0.052	−3640.83	7289.67	2.099	0.045	0.045	0.044
T	3	−2512.22	5032.45	1.860	0.046	0.045	0.045	−3640.93	7291.88	1.863	0.042	0.042	0.042
ST	4	−2511.70	5033.42	1.887	0.041	0.041	0.041	−3639.16	7290.35	1.629	0.044	0.042	0.041
STAB	4	−2526.29	5062.60	2.757	0.060	0.060	0.060	−3645.33	7300.69	2.769	0.060	0.059	0.058

Table 6.4 Goodness-of-fit for the unconditional and conditional case: FX EUR/USD (2002–2008)

Distr.	k	ℓ_N	AIC	$\mathscr{K}$	$\mathscr{AD}_0$	$\mathscr{AD}_1$	$\mathscr{AD}_2$	ℓ_N	AIC	$\mathscr{K}$	$\mathscr{AD}_0$	$\mathscr{AD}_1$	$\mathscr{AD}_2$
NV	2	−1306.09	2618.19	4.117	0.388	0.126	0.116	−2189.33	4384.67	3.307	0.079	0.077	0.075
HS	2	−1293.84	2593.69	1.852	0.070	0.069	0.068	−2193.19	4392.40	2.070	0.087	0.086	0.084
GHS	3	−1290.27	2588.56	1.730	0.058	0.054	0.052	−2183.77	4375.57	2.025	0.054	0.053	0.052
SGHS	4	−1289.87	2589.78	1.897	0.046	0.043	0.042	−2183.34	4376.72	1.896	0.043	0.043	0.042
GSH	3	−1291.00	2590.03	1.642	0.055	0.054	0.053	−2186.16	4380.34	1.591	0.067	0.066	0.065
SGSH1	4	−1290.89	2591.81	1.772	0.051	0.050	0.047	−2185.44	4380.92	1.772	0.056	0.054	0.053
SGSH2	4	−1290.59	2591.22	1.999	0.045	0.043	0.042	−2184.98	4380.00	1.877	0.047	0.046	0.045
BHS	4	−1290.45	2590.93	2.009	0.046	0.044	0.042	−2183.77	4377.57	1.967	0.044	0.043	0.042
EGB2	4	−1290.58	2591.19	2.054	0.045	0.043	0.042	−2183.89	4377.82	1.986	0.044	0.043	0.043
SHS	4	−1293.87	2597.77	1.865	0.069	0.069	0.067	−2193.01	4396.05	2.108	0.081	0.081	0.079
SASHS	4	−1290.45	2590.93	1.873	0.047	0.047	0.045	−2184.23	4378.49	1.838	0.049	0.048	0.048
GH	5	−1288.95	2589.95	1.666	0.047	0.047	0.046	−2185.00	4378.02	2.325	0.057	0.056	0.053
T	3	−1292.66	2593.36	2.142	0.069	0.064	0.063	−2185.53	4381.09	1.881	0.053	0.052	0.051
ST	4	−1292.46	2594.97	2.204	0.061	0.056	0.055	−2182.97	4377.99	1.823	0.046	0.045	0.045
STAB	4	−1299.25	2608.53	2.990	0.066	0.065	0.065	−2188.79	4387.63	2.924	0.065	0.065	0.064

Table 6.5 Goodness-of-fit for the unconditional and conditional case: FX EUR/USD (2008–2012)

Distr.	k	ℓ_N	AIC	$\mathcal{K}$	$\mathcal{AP}_0$	$\mathcal{AP}_1$	$\mathcal{AP}_2$	ℓ_N	AIC	$\mathcal{K}$	$\mathcal{AP}_0$	$\mathcal{AP}_1$	$\mathcal{AP}_2$
NV	2	−1205.17	2416.36	4.198	1.545	0.245	0.220	−1449.77	2905.56	2.831	0.079	0.075	0.073
HS	2	−1174.13	2354.29	1.596	0.049	0.048	0.047	−1444.63	2895.27	2.257	0.070	0.067	0.064
GHS	3	−1174.33	2356.69	2.102	0.056	0.055	0.052	−1439.73	2887.49	1.768	0.039	0.036	0.036
SGHS	4	−1174.03	2356.09	1.494	0.050	0.049	0.048	−1439.73	2889.51	1.774	0.039	0.036	0.036
GSH	3	−1174.02	2358.10	1.505	0.051	0.050	0.049	−1440.25	2888.54	1.646	0.047	0.045	0.042
SGSH1	4	−1174.01	2356.06	1.524	0.050	0.049	0.049	−1440.25	2890.55	1.679	0.046	0.043	0.043
SGSH2	4	−1174.01	2358.08	1.522	0.049	0.049	0.048	−1439.91	2889.87	1.838	0.039	0.038	0.038
BHS	4	−1175.72	2361.50	2.432	0.077	0.061	0.054	−1439.70	2889.45	1.822	0.037	0.037	0.036
EGB2	4	−1174.02	2358.11	1.535	0.051	0.050	0.048	−1439.69	2889.44	1.828	0.037	0.037	0.037
SHS	4	−1174.04	2358.13	1.543	0.051	0.051	0.049	−1444.63	2899.31	2.257	0.070	0.067	0.064
SASHS	4	−1173.94	2357.94	1.733	0.055	0.054	0.053	−1440.40	2890.85	1.821	0.042	0.039	0.038
GH	5	−1174.08	2358.22	1.488	0.050	0.049	0.048	−1439.41	2886.86	1.863	0.038	0.038	0.037
T	3	−1173.45	2358.98	1.546	0.059	0.050	0.046	−1439.39	2888.84	1.926	0.039	0.038	0.038
ST	4	−1174.29	2358.64	2.068	0.052	0.051	0.048	−1439.52	2891.13	1.651	0.041	0.039	0.038
STAB	4	−1179.78	2369.62	3.110	0.065	0.064	0.064	−1440.83	2891.71	2.494	0.052	0.052	0.050

4. **Empirical results**: The results for the unconditional case are summarized on the left side of Tables 6.2, 6.3, 6.4 and 6.5. Across all four examples (and across most of the goodness-of-fit measures), HS dominates NV, BHS dominates EGB2, and SASHS slightly outperforms ST. On the other hand, stable distributions provide a poor fit, whereas GH distributions demonstrate their flexibility although having one additional parameter. $\mathscr{A}\mathscr{D}$-statistics are in general very close for ST, GH, and most of the skewed generalized hyperbolic secant classes.

6.4 Fitting Generalized Hyperbolic Secant Distribution: Conditional Case

1. Integration of GARCH effects: Assuming independent observations—as we did in the last subsection—is not very realistic. To capture dependency between different log-returns, generalized autoregressive conditionally heteroscedastic (GARCH) models have been proposed by Engle [9] and Bollerslev [3] as models for financial return data. These models are able to capture distributional stylized facts (such as thick tails or high peakedness) as well as the time series stylized facts (like volatility clustering). The setting for our GARCH framework is similar to Bollerslev [3] assuming that the log-returns R_t of financial data are given by

$$\Theta_m(L)R_t = \mu + U_t$$

with

$$U_t|F_{t-1} \sim D(0, h_t^2, \eta) \text{ or } U_t|F_{t-1} = h_t\varepsilon_t \text{ with } \varepsilon_t \sim D(0, 1, \eta),$$

where $\Theta_m(L)$ is a polynomial in the lag operator L of order m. For reasons of simplicity, assume that $\Theta_m(L) \equiv 1$ and $\mu \equiv 0$. The residuals $\{U_t\}$ are assumed to follow a GARCH-D process. That means they follow a distribution[3] D with shape parameter η and time-varying variance h_t^2. In the GARCH(1, 1)-Normal specification from Bollerslev [3] h_t^2 is given by

$$h_t^2 = \alpha_0 + \alpha_1 R_{t-1}^2 + \beta_1 h_{t-1}^2 = \alpha_0 + \alpha_1 h_{t-1}^2 \varepsilon_{t-1}^2 + \beta_1 h_{t-1}^2. \tag{6.3}$$

Note that setting $\beta_1 = 0$ results in the ARCH model of Engle [9].

---

[3] Although GARCH models with conditionally normally distributed errors imply unconditionally leptokurtic distributions, there is evidence (see, for example, Bollerslev [4]) that starting with leptokurtic and possibly skewed (conditional) distribution will achieve better results. For that reason, alternative error distributions are used.

2. **Parameter estimation**: We refer to Franke et al. [15], Chap. 12, where ML estimation of standard GARCH models is discussed. In contrast to the unconditional
estimation, where the likelihood L can be easily decomposed as a product of univariate density functions, the main idea is now to represent it as a product of conditional
densities as follows:

$$L(\theta) = \prod_{i=2}^{N} f_D(r_i|R_{i-1} = r_{i-1}, R_1 = r_1; \Theta) f_D(r_1; \Theta).$$

3. **Empirical results**: Time dependencies were taken into account two-fold: First, the
time series were filtered with a suitable GARCH filter and the residuals considered,
instead. Secondly, we directly fitted a GARCH model where the residual distribution
is modeled by each of the distributions mentioned above (see Tables 6.6 and 6.7). In
the first case, especially for the Nikkei data—which are highly skewed—the skew
versions significantly outperform their symmetric counterparts. Again, STAB and
SHS deviate from their competitors (in a negative sense) for most of the goodness-of-
fit measures. In case of the SHS distribution, because of its restriction to highly heavy-
tailed data. Moreover, SASHS and ST are very close together. In the second case, the
GH family achieves the highest likelihood across all four time series, whereas both
the skew GSH versions and the SHS distribution dominate the Anderson–Darling
measures, at least for the Nikkei data set.

To sum up, GH, ST and SGHS, EGB2, BHS, SGSH1 and SGSH2 are very close
together across all measures. In particular, it becomes obvious that most of the generalized hyperbolic secant distributions are fully competitive and, further, we have
interesting statistical properties.

Table 6.6 Goodness-of-fit for GARCH(1,1)-models: NIKKEI and FX EUR/USD (2002–2012)

Distr.	k	ℓ_N	AIC	$\mathcal{K}$	$\mathcal{AD}_0$	$\mathcal{AD}_1$	$\mathcal{AD}_2$	ℓ_N	AIC	$\mathcal{K}$	$\mathcal{AD}_0$	$\mathcal{AD}_1$	$\mathcal{AD}_2$
NV	4	−1362.92	2735.94	23.092	499.9	2.980	0.731	−2410.49	4831.00	18.934	1.369	0.891	0.769
HS	4	−1349.04	2708.18	22.199	0.703	0.694	0.582	−2414.43	4838.88	16.823	0.691	0.599	0.586
GHS	5	−1346.68	2705.49	22.289	0.940	0.728	0.596	−2400.90	4813.82	17.645	0.861	0.685	0.668
SGHS	6	−1341.42	2697.03	26.975	0.643	0.544	0.529	−2398.92	4811.89	18.186	0.887	0.708	0.705
GSH	5	−1345.77	2703.66	22.529	1.284	0.771	0.648	−2399.14	4810.32	17.429	0.836	0.680	0.659
SGSH1	6	−1339.72	2693.63	26.343	0.678	0.637	0.608	−2398.26	4810.56	16.829	0.824	0.674	0.660
SGSH2	6	−1340.77	2695.73	27.742	0.691	0.560	0.559	−2396.13	4806.30	18.698	0.907	0.734	0.721
BHS	6	−1341.36	2696.90	28.887	0.862	0.764	0.610	−2393.86	4801.77	19.132	0.975	0.767	0.737
EGB2	6	−1344.27	2702.72	29.339	1.074	0.778	0.650	−2395.80	4805.64	19.926	1.124	0.844	0.801
SHS	6	−1342.85	2699.89	24.133	0.572	0.493	0.479	−2411.26	4836.55	16.466	0.663	0.578	0.574
SASHS	6	−1341.06	2696.31	25.609	0.859	0.688	0.644	−2397.55	4809.13	17.372	0.950	0.737	0.705
GH	7	−1338.45	2693.14	27.783	0.722	0.583	0.573	−2393.30	4802.65	18.923	0.997	0.774	0.740
T	5	−1345.09	2702.33	22.695	0.800	0.659	0.652	−2398.60	4809.22	18.201	0.979	0.749	0.705
ST	6	−1339.57	2693.34	26.572	0.712	0.579	0.557	−2397.49	4809.01	17.815	0.969	0.747	0.712
STAB	5	−1343.71	2701.62	22.865	0.648	0.601	0.560	−2408.60	4831.23	19.018	0.995	0.741	0.702

Table 6.7 Goodness-of-fit for GARCH(1,1)-models: FX EUR/USD (2002–2008) and FX EUR/USD (2008–2012)

Distr.	k	ℓ_N	AIC	$\mathcal{K}$	$\mathcal{AD}_0$	$\mathcal{AD}_1$	$\mathcal{AD}_2$	ℓ_N	AIC	$\mathcal{K}$	$\mathcal{AD}_0$	$\mathcal{AD}_1$	$\mathcal{AD}_2$
NV	4	−1248.56	2507.15	11.314	1.121	0.737	0.618	−1155.00	2320.06	15.234	0.551	0.526	0.429
HS	4	−1256.73	2523.50	11.935	0.549	0.473	0.472	−1152.33	2314.73	12.783	0.287	0.279	0.277
GHS	5	−1247.29	2506.64	11.815	0.707	0.569	0.539	−1147.07	2306.21	13.291	0.297	0.293	0.290
SGHS	6	−1246.58	2507.24	11.392	0.707	0.572	0.550	−1147.02	2308.15	12.865	0.305	0.297	0.287
GSH	5	−1245.55	2503.15	11.668	0.740	0.598	0.545	−1145.20	2302.48	13.770	0.305	0.302	0.300
SGSH1	6	−1245.19	2504.45	11.561	0.728	0.590	0.558	−1145.32	2304.75	13.292	0.305	0.299	0.295
SGSH2	6	−1244.60	2503.27	11.245	0.742	0.606	0.565	−1145.16	2304.43	13.279	0.313	0.307	0.296
BHS	6	−1244.04	2502.15	10.939	0.739	0.595	0.548	−1144.05	2302.21	13.083	0.321	0.316	0.304
EGB2	6	−1243.48	2501.02	11.672	0.889	0.678	0.608	−1146.22	2306.55	14.103	0.360	0.327	0.323
SHS	6	−1254.58	2523.23	12.239	0.491	0.480	0.472	−1149.03	2312.18	12.269	0.283	0.283	0.276
SASHS	6	−1244.20	2502.47	11.536	0.794	0.618	0.578	−1146.48	2307.08	14.014	0.311	0.306	0.305
GH	7	−1242.65	2501.39	11.152	0.831	0.642	0.580	−1143.82	2303.78	13.008	0.361	0.326	0.315
T	5	−1244.98	2502.01	11.490	0.868	0.657	0.580	−1145.54	2303.16	14.214	0.312	0.305	0.304
ST	6	−1244.47	2503.01	11.284	0.854	0.649	0.591	−1145.51	2305.13	13.993	0.316	0.309	0.301
STAB	5	−1251.30	2516.67	11.378	0.911	0.673	0.615	−1149.46	2313.03	15.478	0.403	0.402	0.327

References

1. Barndorff-Nielsen, O.E.: Exponentially decreasing distributions for the logarithm of particle size. Proc. R. Soc. Lond. A **353**, 401–419 (1977)
2. Barndorff-Nielsen, O.E.: Processes of normal inverse Gaussian type. Finance Stochast. **2**, 41–68 (1998)
3. Bollerslev, T.: Generalized autoregressive conditional heteroskedasticity. J. Econometrics **31**, 302–327 (1986)
4. Bollerslev, T.: A conditional heteroscedastic time series model for speculative price and rate of return. Rev. Econ. Stat. **9**, 542–547 (1987)
5. Brys, G., Hubert, M., Struyf, A.: A robust measure of skewness. J. Comput. Graphical Stat. **13**(4), 996–1017 (2005)
6. Brys, G., Hubert, M., Struyf, A.: Robust measures of tail weight. J. Comput. Stat. Data Anal. **50**(3), 733–759 (2006)
7. Eberlein, E., Keller, U.: Hyperbolic distributions in finance. Bernoulli **3**(1), 281–299 (1995)
8. Elderton, W.P., Johnson, N.L.: Systems of Frequency Curves. Cambridge University Press, London (1969)
9. Engle, R.F.: Autoregressive conditional heteroskedasticity with estimates of the variance of the United Kingdom inflation. Econometrica **50**(4), 987–1007 (1982)
10. Fischer, M.: Skew generalized secant hyperbolic distributions: unconditional and conditional fit to asset returns. Austrian J. Stat. **33**(3), 293–304 (2004)
11. Fischer, M.: A skew generalized secant hyperbolic family. Austrian J. Stat. **35**(4), 437–444 (2006)
12. Fischer, M., Vaughan, D.C.: The Beta-hyperbolic secant (BHS) distribution. Austrian J. Stat. **39**(2), 245–258 (2010)
13. Fischer, M.: A skew and leptokurtic distribution with polynomial tails and characterizing functions in closed form. Working Paper (unpublished). Department of Statistics & Econometrics, FAU Erlangen-Nuremberg, Nuremberg (2012)
14. Fischer, M., Herrmann, K.: The HS-SAS and GSH-SAS distribution as model for unconditional and conditional return distributions. Austrian J. Stat. **42**(1), 33–45 (2013)
15. Franke, J., Härdle, W., Hafner, C.: Statistics of Financial Markets: An Introduction. Springer, Berlin (2004)
16. Gay, D.: Usage summary for selected optimization routines. Computing Science Technical Report No. 153, AT&T Bell Laboratories, Murray Hill, NJ (1990), http://netlib.bell-labs.com/cm/cs/cstr/153.pdf
17. Greene, W.H.: Econometric Analysis. Prentice Hall, Ney Jersey (2003)
18. Lehmann, E., Casella, G.: Theory of Point Estimation. Springer, New York (1998)
19. McDonald, J.B.: Parametric models for partial adaptive estimation with skewed and leptokurtic residuals. Econometrics Lett. **37**, 273–288 (1991)
20. Mittnik, S., Paolella, M.S., Rachev, S.T.: Unconditional and conditional distribution models for the Nikkei index. Asia Pacific Finan. Markets **5**(2), 17–34 (1998)
21. Morris, C.N.: Natural exponential families with quadratic variance functions. Ann. Stat. **10**(1), 65–80 (1982)
22. Nolan, J.P.: Stable Distributions: Models for Heavy-Tailed Data. Birkhäuser, Boston (2005)
23. Prause, K.: The generalized hyperbolic model: estimation, financial derivatives and risk measures. PhD Thesis, University of Freiburg, Freiburg (1999)
24. Zhu, D., Galbraith, J.W.: A generalized asymmetric student t-distribution with application to financial econometrics. J. Econometrics **157**(2), 297–305 (2010)

Appendix A
R-Code: Fitting a BHS Distribution

A.1 Unconditional Case

```
# Data import

temp=scan("C:\\PATH\Nikkei.txt")      # Import exchange rates
data=100*diff(log(temp))              # Calculate log-returns

# Define density function

BHS.density=function(x, SHAPE){
b1=SHAPE[1];b2=SHAPE[2]
return(1/beta(b1,b2)/pi/cosh(x)*((2/pi)*atan(exp(x)))^(b1-1)*
             (1-(2/pi)*atan(exp(x)))^(b2-1))
}

# Define Log-likelihood Function

LOGLIKE=function(PARA,DATA){
mu=PARA[1]; sigma=PARA[2];  shape=PARA[3:4]
ll=-sum(log(1/sigma* BHS.density((DATA-mu)/sigma,shape)))
return(ll)
}

# Start optimization

result=nlminb(start=c(0,1,2,1),obj=LOGLIKE,
             lower=c(-0.1,0,0.01,0.1),upper=c(3,6,10,10),DATA=data)
```

M. J. Fischer, *Generalized Hyperbolic Secant Distributions*,
SpringerBriefs in Statistics, DOI: 10.1007/978-3-642-45138-6,
© The Author(s) 2014

A.2 Conditional Case

```
# Data import

temp=scan("C:\\PATH\Nikkei.txt")      # Import exchange rates
data=100*diff(log(temp))              # Calculate log-returns

# Define density function

BHS.density=function(x, SHAPE){
b1=SHAPE[1];b2=SHAPE[2]
return(1/beta(b1,b2)/pi/cosh(x)*((2/pi)*atan(exp(x)))^(b1-1)*
          (1-(2/pi)*atan(exp(x)))^(b2-1))
}


# Define Log-likelihood Function

LOGLIKE=function(PARA,DATA){
mu=PARA[1];
a0=PARA[2];a1=PARA[3];b1=PARA[4]      # GARCH parameter
shape=PARA[5:6]               # Shape parameter
n=length(DATA);
    h=rep(var(DATA),n);
for(i in 2:n){h[i]=a0+a1*DATA[i-1]^2+b1*h[i-1]}
h=sqrt(h)
ll=-sum(log(1/h*BHS.density((DATA-mu)/h,shape)))
return(ll)
}


# Start optimization

result=nlminb(start=c(0,1,0,0,1.93,1.33),obj=LOGLIKE,
              lower=c(1,0.001,0.001,0.001,0.1,0.98),
              upper=c(1,2,1,1,10,10),DATA=data)
```

Printed by Publishers' Graphics LLC